As of late 2013 there exists an iPad/iPhone application available at the App Store that uses the science-based approach described in this book to make forecasts for stock prices. It is a powerful, user-friendly, and well-documented program capable of making price forecasts with improved accuracy further into the future. Its uniqueness is that it treats stocks like species and the stock market as an ecosystem.

There exist four versions of this application ranging from the full iPad version selling for $49.99 to the light iPhone version selling for $0.99. They all produce price forecasts with quoted uncertainties corresponding to a 95% confidence level.

WATCH A SHORT VIDEO DEMO HERE:
https://www.youtube.com/watch?v=Tqb0ykBHadQ

AN S-SHAPED ADVENTURE

PREDICTIONS – 20 Years Later

Theodore Modis, Ph.D.

Growth Dynamics
Switzerland

Produced by GROWTH DYNAMICS
Via Selva 8, 6900 Massagno, Lugano, Switzerland
Tel. 41-91-9212054
http://www.growth-dynamics.com

ISBN: 978-2-97002-169-8

To Agla

CONTENTS

Contents

PROLOGUE

"Objection sustained! The jury is instructed to ignore these remarks." The judge's stern warning made the defense lawyer shut up and sit down, but as he did he smiled with satisfaction. It is a familiar trick by lawyers in an effort to bias the jury's opinion; the inevitable reaction by the judge that the jury should ignore an unacceptable claim is ineffective. An impression has been created and for all practical purposes cannot be erased from the jurors' memory. No matter how much they try to forget what they heard, something will remain at least in their subconscious mind.

• • •

In a similar situation astute businesspersons think little of the favorite practice among academic forecasters namely to ignore recent-history data and try to predict them with their models. Pragmatic men and women in the marketplace know that once you have seen the outcome it is difficult to forget it. Forecasts can be biased, if unwittingly, in a number of subtle ways. For example, if you already know that a particular product has become a bestseller, you will not persist working with a forecasting model that from early-history sales predicts an imminent decline for the product in question.

The only real proof of success in forecasting involves the test of time. "Make your forecasts and then sit and wait until the time comes," I was told. "Only that way you'll be able to convince people of an enhanced ability to make predictions."

In 1992 Simon & Schuster published my first book *Predictions – Society's Telltale Signature Reveals the Past and Forecasts the Future*.[1] The book broke new ground in understanding society and ourselves by applying fundamental scientific concepts to predicting social phenomena. Twenty years later the challenge arises to confront the predictions made then with the way reality turned out to be. This is something forecasters generally refrain from doing.

It is not the first time that I'll be confronting my forecasts with actual numbers. In 2002 I produced a second edition of *Predictions* with title *Predictions – 10 Years Later.*[2] Ten years of new data were superimposed on the book's original graphs. There were many success stories but also some deviations from the forecasted trends, which proved fertile ground for deeper understanding.

Among the most striking success stories was the diffusion of AIDS in the US. The S-shaped natural-growth curve I had produced with data up to 1988 had indicated a growth process that would be almost complete by 1994 (see Figure 9.9 later in Chapter 9 where the AIDS case is discussed in more detail). In *Predictions* I had concluded that a micro-niche had been reserved for AIDS in American society little more than 1 percent of all deaths. And this was at a time when AIDS had been claiming progressively bigger share of the total number of deaths every year, and some forecasters warned of catastrophic consequences threatening the survival of the human species. But at the end the AIDS "niche" in the US indeed turned out to be far smaller than had been feared by most people. By the late 1990s questions were being raised why forecasts had overestimated the AIDS threat by so much.

Another prediction was confirmed only one month following the book's publication. The production of coal in the UK had been declining along an S-shaped trend for more than twenty-five years but in 1975 a legislative act had fixed production at 125 million tons a year causing a clear deviation from the declining trend. Nine years into this deviation a lengthy miners' strike brought coal production down to the level of the S-shaped trend! But production picked up again after the end of the strike and my prediction was that "something" should happen again to dramatically decrease production. In October 1992, one month after the book's publication, the prediction came true. The miners did not staged another strike, but the government ordered the closing of 61% of the country's mining pits, a move that would bring production down to the level of the forecasted trend. In fact five years later and despite miners' vehement objections coal production in the U.K. was very close to the trend established back in 1975, before any interventions by governments and miners.

The decline in homicides—and criminality in general—during the 1990s had also been predicted in *Predictions* and constitutes another success story. Malcolm Gladwell became rich and famous in 2000 with his bestseller *The Tipping Point* where he likened such "mysterious" sociological changes to those of passing epidemics.[3] But the veil of mystery is largely removed when something can be successfully predicted ahead of time. Moreover Gladwell's explanations of the drop in criminality, such as the broken window theory, have been criticized and disproved.[4]

In *Predictions* I had also ventured some forecasts on the careers of celebrated contemporary personalities who were still producing creatively at the time, three Nobel-prize winners—two physicists and a writer—and a celebrated movie director: Burton Richter (physics 1976), Carlo Rubbia (physics 1986), Gabriel Garcia Marquez (literature 1982), and movie director Federico Fellini. For each one I had determined a productivity curve, which projected forecasts for future works. Indeed ten years later and even twenty years later three of them have pursued their careers in full agreement with the forecasts, (see Appendix Figure 4.5 in Appendix A). In fact, the two physicists have followed their respective curves *very* closely, the writer has followed his curve *quite* closely, but the movie director has not followed it at all. Fellini died in 1993 without making any more movies. For that reason his death must be considered "unnatural". Fellini had most probably been deprived of the chance to complete his life's work. It would be interesting to research into Fellini's later life for plans and scenarios of films that he did not have the time to realize. It sounds ironic but in a way Fellini's death at 73 was more premature than Mozart's at 35 because, as shown in *Predictions*, the latter's curve had been completed by the time of his death, whereas Fellini's curve was interrupted at the level of 74% well below what naturally happens, i.e. at least 90% completion by the time of death, (see the discussion following the figure).

Other deviations from my 1992 forecasts included the evolution of the primary-energy mix and the number of Nobel prizes awarded to Americans. The former, although it generally conforms to a coherent picture has been showing significant deviations from the forecasted trajectories. The latter, has repeatedly proven forecasts to be under-estimating reality. Both subjects involve intricate analyses and have become the object of entire chapters in the book you are now holding, which in that respect is quite different from the original *Predictions* and the follow-up *Predictions – 10 Year Later*.

TWENTY YEARS LATER

Almost all of the 81 graphs in *Predictions* are updated once again here and there is a discussion on the way reality behaved in the last twenty years with respect to the way it had been expected. A few graphs have not been updated either because the publication of the data series in question has been discontinued, for example time-series data for the use of horsepower, or because the growth process has been completed and there cannot be more data on it, for example the extinction of steam-engine locomotives.

But contrary to *Predictions – 10 Years Later* the book you are now reading is not a new edition of *Predictions*. There is a brief introduction to S-curves so that the reader is not required to have read *Predictions*, but a large part of this book is devoted to new material, namely a selection of natural-law applications (S-curves and related formulations) among the many I have studied during the last twenty years.

The reader will also find here the primordial growth curve that describes how change and complexity appeared in the Universe from the very beginning—the Big Bang—to today and beyond. When I first gathered together the data for this curve in 1995 I unwittingly contributed to the then emerging Singularity movement. For the sake of the readers not familiar with this movement let me say that Singularitarians are an eccentric group of people who in the name of science—often misused—predict that by 2045 machines will take over society reducing human beings to second-class citizens. They come to this conclusion drawing in part on my data for the primordial curve while my own conclusions from the same data are quite the opposite, namely that we presently are witnessing the maximum rate of change and that the lives of our children and our grandchildren will progressively become less turbulent.

This optimism on my part is neither naïve nor religiously motivated as some critics have insinuated. It follows my understanding of the law of growth in competition. A natural-growth process may depict an explosive character early on but only for a limited period of time. Invariably its rate of growth slows down and the process generally proceeds to completion. No niche in nature was ever left partially filled (or emptied in case of niche depletion) under normal circumstances. This last condition "under normal circumstances" has also been criticized with the argument that in an ever-changing environment circumstances appear as exceptional, i.e. not normal, more often than not. But the definition of normal in this context incorporates all those kinds of things that took place during the historical window during which data were collected for the determination of the growth curve. For example, while studying a product's life cycle normal conditions include the appearance of innovative competitive products in the market. But the appearance of a world war should not be considered normal for a study of data spanning the last fifty years. Similarly, a nervous breakdown or a suicide is expected to introduce important deviations from the smooth curve describing the evolution of an artist's life work. Such disagreements constitute intriguing telltale signals. Predictions that come out wrong when real-life trajectories deviate from natural-growth patterns may hide secrets. I have found it rewarding to dig deeper into the reasons that cause deviations from fundamental natural laws. The insights obtained this way reveal as much if not more than successful forecasts do.

WHEN REAL LIFE MISBEHAVES

Growth in competition is a simple and powerful law. We know from physics that the simpler a law, the more fundamental it is and the wider its range of applications. But it is not the only law in effect and its simple formulation is often encumbered with complications that cause deviations from natural trends. One complication stems from mutations, which are more frequent in the marketplace than in nature. Another stems from the opening of a new niche after the old one has been filled. Another complication stems from the simultaneous presence of more than one "species" in the same niche. The presence of lynxes has a catastrophic impact on the evolution of a rabbits population, which otherwise would have followed an ideal S-shaped curve. Two species can impact each other's rate of growth in negative, positive, or neutral ways yielding six combinations of coexistence (predator-prey, symbiosis, parasitism, etc.) Three species in the same niche yield even more complicated situations. So there are many reasons for deviations from simple natural-growth patterns and for forecasts to be proven wrong. Some situations are so entangled that unraveling the driving forces becomes imponderable. A case at point is the fate of my first book *Predictions*.

The manuscript had from the very beginning been deemed a promising bestseller. Simon & Schuster had imposed Herculean tasks on me before formally taking on the project. Because my English "smelled" foreign they asked me to find an American-born science writer who would write the book for me (they had specified someone like a *Scientific American* writer, or the editor of the science column in *Time* magazine).

They also wanted written reassurances from three world-renowned scientists affirming that the content of the manuscript is scientifically sound. Then the vice-president editor that was coaching me made it clear that one couldn't take a shot at a bestseller without a household science-name endorsement on the cover from such people as Asimov or Carl Sagan. "A Nobel Laureate would do, even if his/her name is not well known," he conceded. Finally, and for my own sake this time, I should have at least three *relevant* people go over the manuscript meticulously looking for inconsistencies and errors.

In pursuit of these aims I came in contact with individuals and became entangled in situations that ranged from grotesque to impossible so much so that I decided to recount all of my experiences in a little book entitled *Bestseller Driven*. However, at the end, with perseverance and luck I managed to satisfy all requirements imposed on me with flying colors and the publisher dished out a handsome bestseller's advance. But the market did not react as expected. The book sales did not reach the volumes Simon & Schuster had hoped for and there was no easy explanation for it.

When forecasts disagree with reality it is generally because they are not good, but when well-done forecasts perform poorly, it may not be their fault. Being in the forecasting business I have sometimes witnessed exceptional forecasts endowed with high confidence levels become disproved by data that no longer behave in a manner consistent with past behavior. In these cases, I am convinced it is not the forecaster's fault but that reality simply "misbehaves."

It is indeed possible that reality might misbehave, and science is not immune to such an eventuality. Experimental physicists refrain from publishing discoveries based on less than five standard deviations. A five-standard-deviation discovery carries the remote possibility that once in two million times the observation will be consequence of a statistical fluctuation rather than a real phenomenon. Obviously, the more accurate the measurement, the smaller this possibility will be. But even the most venerated discoveries—like those leading to a Nobel Prize award—often carry a very small probability of not being true. If it becomes realized, we should not blame the researcher, or the experimental methods used, or the Nobel committee. When such a thing happens, it is just a matter of bad luck!

My motivation and confidence in confronting the predictions I made twenty years ago with reality stem from the forecasting techniques I used, which differ significantly from those of most forecasters because mine rely on natural laws. In particular, the law of natural growth in competition—otherwise known as survival of the fittest or Darwinian competition—gives rise to the ubiquitous S-shaped curve, which enters our lives in so many ways. It generally makes it possible to see clearer further into the future. It is neither magical nor trivial but leads to forecasts that have the blessing of science and are less vulnerable to subjective biases, such as wishful thinking.

1

The S-Curve

"Easy come easy go", "Early ripe, early rot", "When it rains it pours", "You need money to make money", "All beginnings are difficult".

• • •

Proverbs like these are products of popular wisdom but they can be rigorously derived from the mathematical formulation of the S-curve.

THE LAW OF NATURAL GROWTH

At the heart of competition lies the principle of survival of the fittest. If you put a pair of rabbits in a meadow and the average rabbit litter is taken as two, you can watch the rabbit population go through the successive stages of 2, 4, 8, 16, 32, 64, ..., 2^n in an exponential growth. There is a population explosion up to the time when a sizable part of the ecological niche is occupied. It is only after this time that limited food resources begin imposing constraints on the number of surviving rabbits and the population growth slows down as it approaches a ceiling—the capacity of a species' ecological niche. This slowdown may happen by means of increased kit mortality, diseases, lethal fights between overcrowded rabbits, or even other more subtle forms of behavior that rabbits may act out unsuspectingly. Nature imposes population controls as needed, and in a competitive environment, only the fittest survive.

Over time, the rabbit population traces an S-shaped trajectory, see Figure 1.1. The *rate* of growth traces a curve that is bell-shaped and peaks when half the niche is filled. The S-shaped curve (S-curve) for the population and the bell-shaped curve for its rate of growth constitute a pictorial representation of the natural growth process—that is, how a species population grows into a limited space by obeying the law of survival of the fittest.

At the ceiling, we may witness oscillations as the rabbit population explores the possibility to go further and overshoots the niche capacity only to fall back later giving the grass a chance to grow back and feed more rabbits. At this point we may talk of a *homeostasis*, a stable state of equilibrium between the number of rabbits and the amount of grass.

An S-curve and the associated life cycle are two different ways of looking at the same growth process. The S-curve represents the size of the growth and points out (anticipates) the growth potential, the level of

Natural Growth in Competition and Its Life Cycle

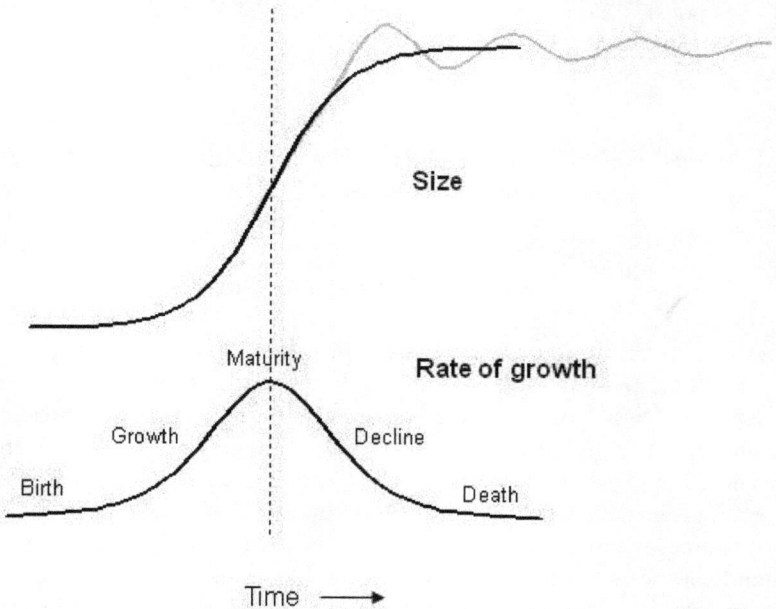

FIGURE 1.1 The S-shaped pattern above depicts the natural-growth process. The bell-shaped curve below depicts the rate of growth and is used across disciplines as a template for the cycle of life. The gray line sketches the fluctuations of a rabbit population that has filled its ecological niche to capacity.

16

the final ceiling, how much could one expect to accomplish. The bell-shaped life-cycle curve represents the rate of growth and is more helpful when it comes to appreciating the growth phase you are traversing, and how far you are from the end. The S-shaped curve reminds us of the fact that competitive growth is capped. The bell-shaped curve reminds us that whatever gets born eventually dies. From an intuitive point of view, an S-curve promises a certain amount of growth that can be accomplished, whereas a bell-curve heralds the coming end of the process as a whole. Both curves possess predictive power.

There is predictability also during the rapid-growth phase (*rheostasis*). You can easily anticipate where a fast-moving train will end up. A popular business metaphor is the supertanker whose imminent course can be trivially predicted. A bicycle is stable only when in motion and the faster it is going the more stable it is and the easier it is to project its trajectory.

The predictive power of the bell-shaped life-cycle curve comes from its symmetry. A rapid rise will be followed by an equally rapid decline, echoing such expressions as "Easy come, easy go" and "Early ripe, early rot". Many business endeavors have experienced this the hard way in the marketplace.

The mathematical equation—the logistic equation—that describes the law of natural growth in competition and gives rise to the S-curve says in words that the rate of growth must be at all times proportional to two things:

- The amount of growth *already accomplished.*
- The amount of growth *remaining to be accomplished.*

If either one of these quantities is small, the rate of growth will be small. This is the case at the beginning and at the end of the process. The rate is greatest in the middle, where both the growth accomplished and the growth remaining are sizable. Furthermore, growth "remaining to be accomplished" implies a limit, a saturation level, a finite niche size. Competition is a consequence of a limited resource and therefore growth in competition cannot go on forever; it is necessarily capped. This ceiling of growth is assumed to be constant throughout the growth process. Such an assumption is a good approximation to many natural-growth processes, for example, plant growth, in which the final height is genetically pre-coded.

It is a remarkably simple and fundamental law. Besides used by biologists to describe species populations, it has also been used in medicine to describe the diffusion of epidemic diseases. J. C. Fisher and R. H. Pry referred to the logistic equation as a diffusion model and used it to quantify the spreading of new technologies into society.[1] One can immediately see how ideas or rumors may spread according to this law. Whether it is ideas, rumors, technologies, or diseases, the rate of new occurrences will always be proportional to how many people have it and to how many don't yet have it. At the end you will always be able to find—albeit in slowly diminishing numbers—the outcasts who never heard the rumor, or refused to adopt the new technology.

The S-curve has also being referred to as a learning curve in psychology as well as in industry. For example, the evolution of an infant's vocabulary has been shown to follow an S-curve that reaches a ceiling of about 2500 words by age six.[*] Acquiring vocabulary can be thought of as a competitive process where words in the combined active vocabulary of the two parents compete for the infant's attention. The words most frequently used will be learned first, but the rate of learning will eventually slow down because there are fewer words left to learn. This ceiling of 2500 words defines the size of the home vocabulary "niche," all the words available at home. Later, of course, schooling enriches the child's vocabulary, but this is a new process, starting another cycle, following probably a similar type of curve to reach a higher plateau.

The S-curve (black line) in Figure 1.1 is asymptotic, i.e. it approaches zero, and the level of the ceiling continuously but reaches these values only at times $-\infty$, $+\infty$ respectively. On the other hand the fact that growth is proportional to the amount of growth already achieved renders the beginning of every natural-growth process practically very difficult (theoretically impossible because zero growth achieved yields a null rate for growth and so things cannot be started!) This demystifies the known difficulty associated with beginnings. An ancient Greek proverb on achievement equates the beginning with half of the whole! The consequences on learning are enlightening. Theoretically, learning cannot begin without outside help. The work of teachers becomes indispensable in this context. The teacher is the custodian of knowledge and oriental schools of thought preclude search for esoteric knowledge and personal development without a teacher.

The theoretical difficulty in getting growth in competition started touches upon philosophical questions akin to the genesis because of the requirement that some discontinuous intervention from an external

---

[*] An S-curve has been fitted on the data found in Whiston (1974).[2]

agent (for example, a powerful intelligent entity) is necessary in order to get something going from nothing.

FATAL CAR ACCIDENTS

The logistic equation has been successfully used to describe growth processes where the notion of competition has been raised to remarkable levels of abstraction. Cesare Marchetti has argued that there is competition among primary-energy sources for consumers' favor and that there is competition among diseases for victims.[*] In all cases there is a limited resource, which imposes the constraint that only the best-fit candidate will win. My favorite example is fatal car accidents.[3] All possible accidents can be thought to compete for becoming realized and claim victims. Only the "best" of them will do so because here again there is a limited resource and contrary to what one may naively expect this limited resource is much smaller than the entire population.

Car safety has been a passionate subject frequently appearing in headlines. At some point in time cars had been compared to murder weapons. Still today an estimated 1.2 million people worldwide die from car accidents every year, and 50 million suffer injuries. Efforts are continually made to render cars safer and drivers more cautious. How successful have such efforts been? Can this rate be significantly reduced as we move toward a more advanced society?

To answer these questions, we must look at the history of car accidents, but in order to search for a fundamental law we must have accurate data and an appropriate indicator. Deaths are recorded and interpreted with less ambiguity than other accidents. Moreover, the car as a public menace is a threat to society, which may "feel" the pain and react accordingly. Consequently, the number of deaths per one hundred thousand inhabitants per year becomes a better indicator than accidents per mile, or per car, or per hour of driving.

The data shown in Figure 2 are for the United States starting at the beginning of the 20th century. What we observe is that deaths caused by car accidents grew along an S-curve with the appearance of cars until the mid 1920s, when they reached about twenty-four per one hundred thousand per year. From then onward they seem to have stabilized, even

---

[*] Working at IIASA (International Institute of Applied Systems Analyses) in Luxemburg, Vienna, Cesare Marchetti was the first to employ S-curves extensively in the widest range of applications.

though the number of cars continued to grow. A homeostatic mechanism seems to emerge when this limit is reached, resulting in an oscillating pattern around the equilibrium position. The peaks may have produced public outcries for safety, while the valleys could have contributed to the relaxation of speed limits and safety regulations. What is remarkable is that for over sixty years there has been a persistent self-regulation on car safety despite major increases in car numbers and performance, and important changes in speed limits, safety technology, driving legislation, and education.

Why the number of deaths is maintained constant and how society can detect excursions away from this level? Is it conceivable that someday car safety will improve so much that car accidents will be reduced to zero? American society has tolerated this level of accidents for more than half a century. A Rand analyst has described it as follows: "I am sure that there is, in effect, a desirable level of automobile accidents—desirable, that is, from a broad point of view, in the sense that it is a necessary concomitant of things of greater value to society".[4] Abolishing cars from the roads would certainly eliminate car accidents, but at the same time it would introduce more serious hardship to citizens.

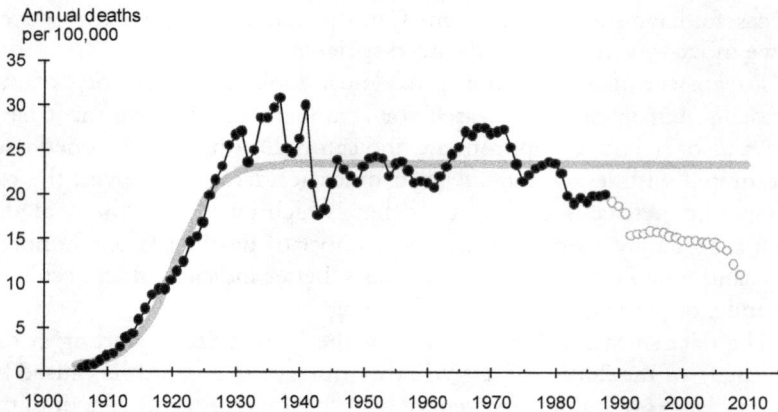

Fatal Car Accidents in the US

FIGURE 1.2 The annual number of deaths from motor-vehicle accidents per 100,000 population has followed an S-curve to reach a ceiling of 24 around which it has been fluctuating since the mid 1920s, not unlike the rabbit population sketched in gray in Figure 1.1. The peak in the late 1960s provoked a public outcry that resulted in legislation making seat belts mandatory. The light-color open circles indicate data during the last 20 years.

An invariant homeostatic level can be thought of as a state of well-being. It has its roots in nature, which develops ways of maintaining it. Individuals may come forward from time to time as advocates of an apparently well-justified cause. What they do not suspect is that they may be acting as agents to deeply rooted forces maintaining a balance that would have been maintained in any case. An example is Ralph Nader's crusade for car safety, *Unsafe at Any Speed*, published in 1965, by which time the number of fatal car accidents had already demonstrated a forty-year-long period of relative stability.[6] But examining Figure 1.2 more closely, we see that the mid 1960s show a small peak in accidents, which must have been what prompted Nader to blow the whistle. Had he not done it, someone else would have. Alternatively, a timely social mechanism might have produced the same result; for example, an "accidental" discovery of an effective new car-safety feature.

During the last thirty years there has been evidence for a gentle downward trend shown with little circles in Figure 1.2. One could argue that Nader's crusade for car safety was indeed effective. After all it was instrumental in making seat belts mandatory and lowering the speed limits throughout the country. I seriously doubt such cause-and-effect reasoning. Seat belts and speed limits certainly had some effect, which among other things, made environmentalists shift their focus to other issues. But action taken forty years ago would not still keep reducing deaths from car accidents today. In fact speed limits have been significantly raised in most states since then. The state of Montana has even experimented with lifting some speed limits altogether.

As usually, there is a more deeply seated explanation for deviations from a natural-growth pattern. The airplane has been steadily replacing the automobile as a means of intercity transportation since the late 1960s. As we will see in Chapter 3 despite the fact that the automobile still commands a dominant share of the transportation market today, Americans have in fact been giving up, slowly but steadily, their beloved cars, and the fatal accidents that go with them. The recent data in Figure 1.2 do not disprove the forecast but illustrate the case when a new phenomenon—e.g. the natural substitution of airplanes for cars—introduces a deviation from the projection of a natural path. The gaining of such understandings was among the motivations for writing this book.

INVARIANTS

Fatal car accidents exemplify the concept of an *invariant*, something that does not change over time and location. Invariants are, of course, the easiest things to forecast. They reflect states of equilibrium maintained by natural regulating mechanisms. They often are the ceiling of an S-curve as shown in Figure 1.1. In ecosystems such equilibrium is called *homeostasis* and refers to the harmonious coexistence of predator and prey in a world where species generally do not become extinct for natural reasons.

States of equilibrium can also be found in many aspects of social living. Whenever the level of a hardship or a menace increases beyond the tolerable threshold, corrective mechanisms are automatically triggered to lower it. On the other hand, if the threat accidentally falls below the tolerated level, society becomes blasé about the issue, and the corresponding indicators begin creeping up again with time.

In *Predictions* I discussed many invariants found in social situations. For example human beings around the world are happiest when they are on the move for an average of about seventy minutes per day. Going above or below this norm causes unhappiness and discomfort, and is met with aversion and rejection. To forget the fact that one is moving for longer periods, trains feature reading lounges and bar parlors. Airlines show movies during long flights. On the other hand, lack of movement is equally objectionable. Prisoners pace their cells back and forth in order to meet this "quota" of travel time.

During these seventy minutes of travel time, people like to spend around 15 percent of their income on the means of travel? To translate this into biological terms, one must think of income as the social equivalent for energy. And these two conditions are satisfied in such a way as to maximize the distance traveled? Poor people walk, those better off drive, while the rich fly. From primitive African tribes to sophisticated metropolitan residents, they are all trying to get as far as possible within the seventy minutes and the 15 percent budget allocation. Affluence offers a bigger radius of action. Fast airplanes did not shorten travel time; they increased the distance traveled.

Maximizing range is what all living organisms are trying to achieve from the most primitive ones to humans. The rapid multiplication of unicellular amoebas but also space exploration both aim at expansion into space as far as possible; as Marchetti put it, every other rationalization tends to be poetry.

There are several invariants associated with the use of cars. The average car speed in the United States has not really changed since

22

Henry Ford's time. Ever since the automobile appeared in America drivers seem to have been confined to an average speed of 30 miles per hour. All road and car-performance improvements during the 20th century have mostly compensated for the time lost in traffic jams and at traffic lights.

At an average speed of 30 miles per hour, a traveling time of seventy minutes a day—the natural invariant for the daily displacement—translates to about 35 miles. This daily quota for car mileage is indeed corroborated from data on car statistics. For more than half of the 20th century annual mileage in the United States was confined to around 9,500 miles, despite the great advances in car speed and acceleration over this period. This turns out to be 36.4 miles per working day, in good agreement with the daily displacement quota of 35 miles.

Invariants have the tendency to hide behind headlines. For example, the number of deaths due to motor vehicle accidents becomes alarming when reported for a big country like the United States over a three-day weekend, which is what journalists do. However, when averaged over a year and divided by one hundred thousand of population, it becomes stable over time and geography.

In 1995 I was invited to address medical doctors in Geneva attending their annual meeting. Following my talk there was a lively discussion on the action of natural laws and in particular invariants. The next speaker talked about in-hospital infections and how difficult it proved lowering them below the level of 25 percent despite coordinated efforts on many fronts: hygiene, organization, isolation, decontamination, etc. At the end he concluded that it might be counterproductive to insist trying to lower infections further because this level may represent an invariant, a "desirable" homeostatic level in the same sense as deaths from car accidents. In-hospital infections may be concomitant of things of greater value to the hospital, such as being able to treat many different diseases simultaneously. Obviously if the hospital admitted patients who suffered from only one disease, e.g. cirrhosis of the liver, or tuberculosis, in-hospital infections could be greatly reduced. The price for being able to treat all diseases in one place is a certain "quota" of in-hospital infections just as the price of having the use of cars was to tolerate a certain "quota" of fatal car accidents.

DEVIATIONS

Apart from the introduction of a new natural phenomenon—such as the substitution of airplanes for cars—there can be many other reasons why the evolution of a natural-growth process may deviate from an S-curve. Unnatural events, such as a world war, major earthquakes, natural or economic catastrophes, and any other singular but significant phenomenon that never occurred before during the historical window under study, may introduce short-term deviations from the S-curve trend. But when the unnatural event subsides, a return to the S-curve course is expected. Other deviations from S-curves may involve simple fluctuations (statistical in nature or otherwise), and such phenomena as cascading S-curves, the early catching-up effect, the final overshoot, a niche within a niche, and a niche with variable capacity. Most of these will be discussed in detail in Chapter 5, but let me give you here a flavor of what such deviations may look like.

It was shown in *Predictions* that in contrast to Mozart's textbook-like S-curve, Robert Schumann's curve was punctuated by a nervous breakdown and a suicide attempt, both considered "unnatural" events. The evolution of Schumann's published compositions may seem to roughly follow an overall S-shaped pattern but at closer look, one distinguishes three smaller S-curves, one up to his nervous breakdown,

Robert Schuman (1810-1856)

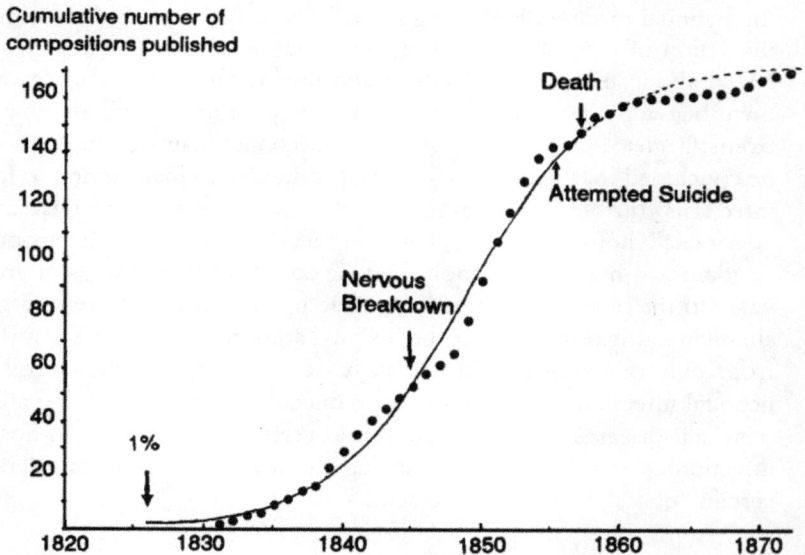

FIGURE 1.3 The publication of Schumann's compositions. The fitted curve begins around 1826 and aims at a ceiling of 173. Schumann's publications reached 170, sixteen years after his death.

24

another one up to his attempted suicide, and a third one that includes compositions published posthumously, see Figure 1.3, which shows the cumulative publication of Schumann's compositions.[6] They oscillate somewhat around the fitted trend. The overall curve is segmented into three periods by his major nervous breakdown in 1845 and his attempted suicide in 1854. Both of these events interfered with the natural evolution of his work, yet one can distinguish a smaller S-curve for each period.

In another situation the evolution of Alfred Hitchcock's life work as a movie director shows a deviation early on from an otherwise exemplary S-curve. In fact, Hitchcock's S-curve seems to originate well before 1925 when his first movie appeared. This means that his impulse for direction was deeply rooted. But for "technical" reasons (one cannot reasonably expect a teenager to succeed in undertaking the direction of full-length features) his work lagged behind the theoretical productivity curve in the beginning. When he finally started his career as a film director at twenty-six, he produced prodigiously during the first six years, as if he were trying to "catch up," see Figure 1.4. Such catching up—displaying a sudden release of pent-up productivity—is common among people whose lifework mainly consists of being bosses or directors because no one begins a career at that level. It is also clearly visible in Fellini's life work mentioned earlier, (see Appendix Figure 4.5 in Appendix A).

Hitchcock's curve also demonstrates the niche-within-a niche effect. From 1930 onwards the cumulative number of Hitchcock's full-length features grows smoothly to reach fifty-two by 1975, 96 percent of his potential specified by the curve's ceiling at fifty-four. Here is another case in which little significant productive potential remained to be realized at the time of death. The twist in Hitchcock's case is that in 1955 he was persuaded to make television films for the celebrated series Alfred Hitchcock Presents. The open circles on the graph represent the sum of both the full-length and the shorter television films. A smaller S-curve can be clearly outlined on top of the large one. This niche-within-a-niche contains twenty films; the process of filling it up starts in 1955, and flattens out approaching natural completion by 1962.

The evolution of Hitchcock's work just before he embarked on the television adventure contains a suggestive signal, a slowing down leading smoothly into the television activity that follows. Statistically speaking, the small deviation of the data points around 1951 has no real significance. It coincides, however, with the times when the film industry in the United States felt most strongly the competition from the growing popularity of television.

Alfred Hitchcock's Two Niches in Cinema

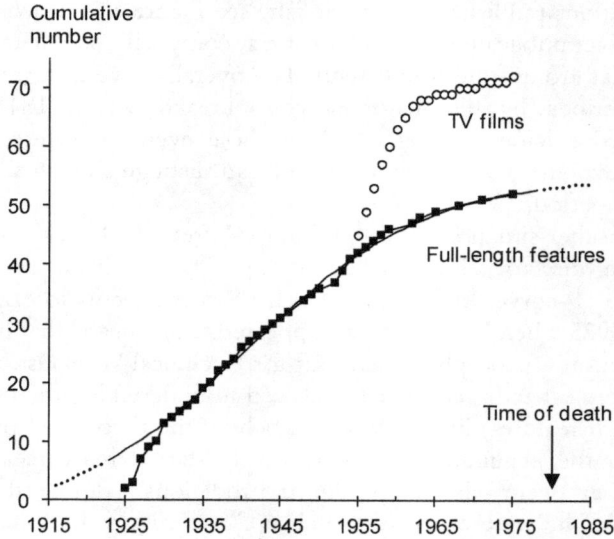

FIGURE 1.4 The squares indicate full-length films while the circles indicate the sum of both full-length plus shorter television films. The fit is only to the full-length films. A smaller curve is outlined by the television works and seems to have its beginning in Hitchcock's film works. The configuration of these two S-curves provides a visual representation of the niche-within-a-niche situation.[7]

Another frequent deviation appears as we approach the ceiling of the S-curve. In Figure 1.1 we saw an oscillatory behavior around the ceiling of the S-curve as the species population overshoots the niche capacity only to fall back later below the homeostatic level. This oscillation may seem regular but in fact results from a random search of the species exploring possibilities for growth. Mathematically when the rate of growth drops to zero (at the end and the beginning of the S-curve) chaotic fluctuations appear. Alain Debecker and I obtained a mathematical understanding of this phenomenon by linking S-curves to chaos in a way different from the classical approach described in the bestseller *Chaos*.[8]

Finally, the niche capacity may change over time, which will result in a deviation from the classical S-shaped pattern. This was the case with the number of US Nobel laureates the evolution of which seemed to have covered more than half of its S-curve by 1988. The forecast in *Predictions* had the annual rate of American Nobel laureates on a declining trend and yet twenty years later no decline in this number has been observed. We will analyze this case in detail in Chapter 8, but here it suffices to say that the wrong forecast was due to the assumption that the niche capacity for American Nobel laureates is constant over time. This niche has been increasing and not only because the US population is increasing. America, as a rule, welcomes research scientists from all over the world while it thwarts immigration by the uneducated. It makes sense that the population sample capable of producing Nobel laureates in America has been growing faster than the rest of the population.

Deviations from S-curves may trigger more intricate elaborations of the natural-growth approach such as having a niche with variable capacity or having more than one species in the niche, which has been formulated as the Volterra-Lotka model—fully discussed in Chapter 8. These refinements can shed more light in special cases, but they do not diminish the usefulness of the basic approach. As a first approximation the visualization of natural growth in competition can be done either with the bell-shaped or with the S-shaped curve. The important difference between them, from the statistical analysis point of view, is that the S-curve, which usually depicts a cumulative rate of growth, is much less sensitive to fluctuations because one year's low is compensated for by another year's high. Therefore, the S-curve representation of a natural growth process provides a more reliable way to forecast the level of the ceiling. From an intuitive point of view, an S-curve promises the amount of growth that can be accomplished while a bell-curve heralds the coming end of the process as a whole.

2

Limits to Growth

"Most persons think that a state in order to be happy ought to be large; but even if they are right, they have no idea what is a large and what a small state...To the size of states there is a limit, as there is to other things, plants, animals, implements; for none of these retain their natural power when they are too large or too small, but they either wholly lose their nature, or are spoiled?"

Aristotle 350 B.C.

• • •

In 1972, a book appeared under the title, *The Limits to Growth*, published by the Club of Rome, an informal international association with about seventy members of twenty-five nationalities.[1] Scientists, educators, economists, humanists, industrialists, they were all united by their conviction that the major problems of mankind today are too complex to be tackled by traditional institutions and policies. Their book drew alarming conclusions concerning earth's rampant overpopulation and the depletion of raw materials and primary energy sources. Its message delivered a shock and contributed to the "think small" cultural wave of the 1970s, which nevertheless came and went like a fad leaving behind it a rather insignificant permanent mark. Today business mottos like "bigness is goodness" seem to be in effect as much as ever.

Society cannot forget the long decades of "fat cows" that followed World War II and growth is still today widely considered indispensable for prosperity. In particular the growth of Gross Domestic Product (GDP) is considered to be an essential ingredient of a healthy economy.

A plethora of near-future GDP forecasts typically project a constant percentage over several years in the future invariably promising growth.[2] But longer-range forecasts are also often based on more or less exponentially growing patterns.[3] And yet there have been voices advocating that days of diminishing growth are approaching. These voices began in 1972 with the publication of *The Limits to Growth by the Club of Rome*, but they increased in numbers recently with such works as Tim Jackson's *Prosperity Without Growth*,[4] Serge Latouche's *Farewell to Growth*,[5] and Peter Victor's *Managing Without Growth: Slower by Design , Not Disaster*.[6] Richard Heinberg has a rather extensive compilation of publications on this subject in his book *The End of Growth*.[7]

Complementary to the works mentioned, which are generally based on economic arguments, I want to address the growth of GDP as a natural-growth process.[8] Growth in competition is an appropriate eyepiece here because there is abundant competition in the processes that contribute to the formation of GDP and the issue of limited resources cannot be denied. Competition and limited resources are the ingredients of natural growth that describes how species populate ecological niches. But my ultimate argument for using an S-curve approach is an *a posteriori* one, namely the goodness of the way an S-curve describes the evolution of GDP over almost a century.

Natural growth implies a cap, a final ceiling, in sharp contradiction to forecasts based on linear and exponential patterns, which are unlimited. I will present actual data on GDP growth demonstrating that such a cap is altogether realistic. Once at the ceiling there can be no more growth, none unless catastrophes and disasters of unseen-before magnitude create new niches for growth or the "species" undergoes a major mutation effectively transforming itself into a different species, e.g. through war and conquest of new territory.

GDP GROWTH IS CAPPED

The historical data on the US GDP come from the US Department of Commerce, Bureau of Economic Analysis, and the US Census Bureau.[9] They consist of yearly data up to the end of 2012. Faced with the dilemma of studying the nominal GDP expressed in current dollars or the real GDP—i.e. corrected for inflation—expressed in constant (chained) dollars, I decided to try them both. To my surprise both sets of data result in excellent S-curve fits. Figure 2.1 shows nominal GDP per capita and Figure 2.2 shows real GDP per capita. The lower graphs show the rate of change in annual increments, i.e. the life cycle of each process; they are derived from the curves at the top of Figures 2.1 and 2.2. The emerging images indicate that the inflection points—centers of the life-cycle curves—are behind us, particularly for nominal, and that there is about a 7-year lead by nominal GDP.

The goodness of the fits can be visually appreciated by the way the data points closely follow the S-curve patterns (thick gray lines) over 80 years despite many world-shaking events and varying inflation over this period of time.

A further surprise is that whereas the nominal GDP nears completion of the growth process, the curve is 77.6% completed by the end of 2012, the real GDP has still considerable remaining growth potential, its curve being only 55.2% completed. The respective midpoints—inflection points of the S-curves—were in mid 1998 for nominal GDP and late 2005 for real GDP. Thin black lines delimit 90% confidence-level bands, in other words, where we should expect future GDP values to fall nine times out of ten. Such bands are established using look-up tables from an extensive study on uncertainties we carried out with Alain Debecker.[10]

The open circles shown for the period 2013-2017 are forecasts by the International Monetary Fund (IMF) made in traditional economists' ways, most frequently consisting of linear or exponential extrapolations.[11] For nominal GDP they indicate a more optimistic trend progressively deviating from the S-curve course (outside the 90% confidence-level band). The big black dot in Figure 2 is a long-term forecast by Agnus Maddison.[12] There is a rather limited amount of growth potential ahead of us in nominal GDP but a significant amount in real GDP, which economists prefer to talk about more often than not. But what good could it be to know that hidden somewhere behind the numbers there is much growth to come while all we see in everyday life in current dollars is little growth? The real GDP seems to have little relevance to the people on the streets. Apparently inflation is "eating up" all our growth potential. But given that both nominal and real GDP constitute natural-growth processes—i.e. follow S-curves—inflation, which links the two must also have some natural origin rather than its usual attribution to frivolous human behavior!

US Nominal GDP per Capita in Current dollars

Thousand $

Annual Increments

Thousand $

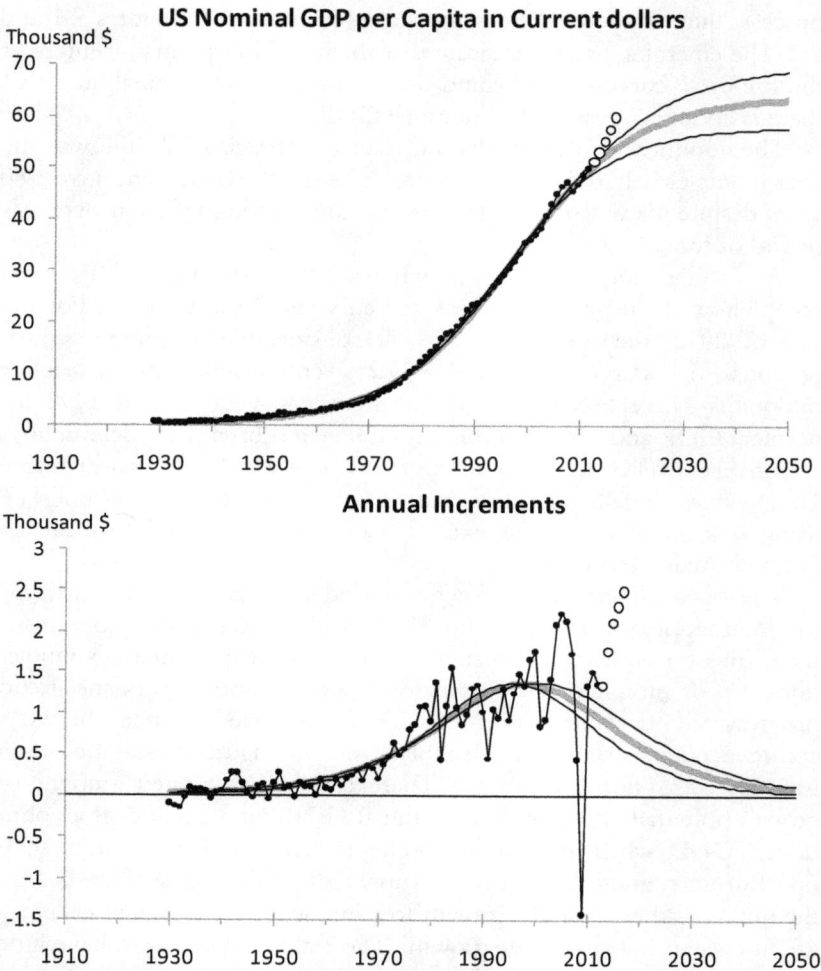

FIGURE 2.1 Above, nominal GDP per capita (black dots) and S-curve fit (thick gray line). Below, the rate of growth in annual increments. The open circles are IMF forecasts. The thin black lines delimit 90% confidence-level bands.

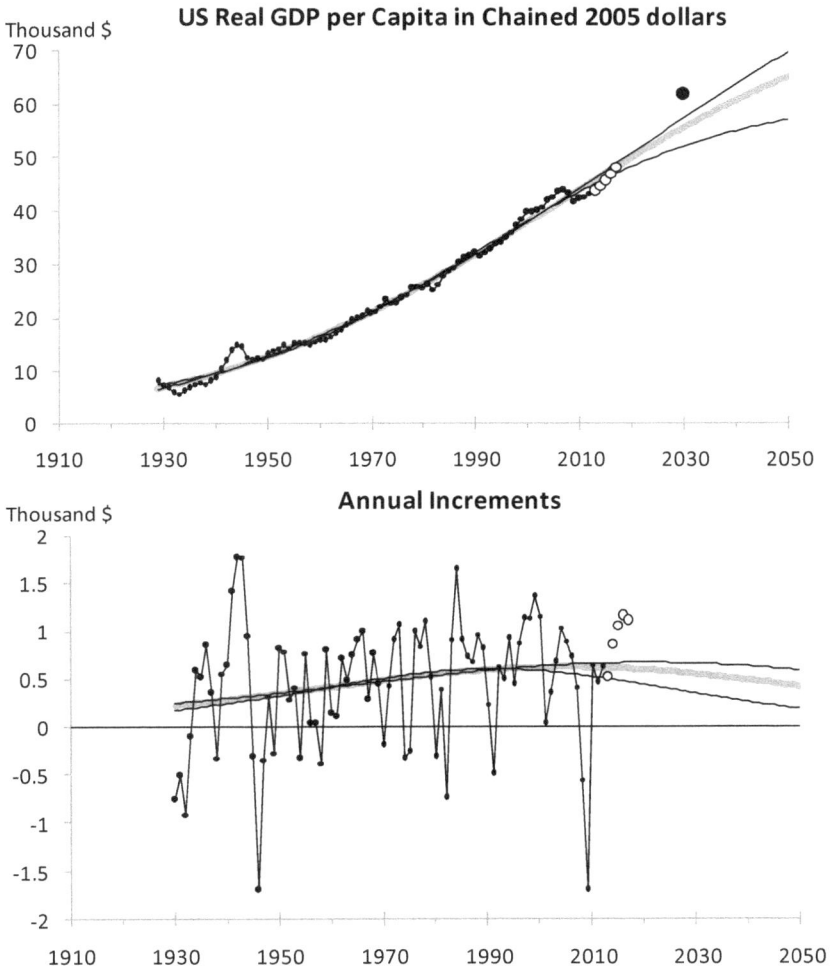

FIGURE 2.2 Above, real GDP per capita (black dots) and S-curve fit (thick gray line). Below, the rate of growth in annual increments. The open circles are IMF forecasts. The thin black lines delimit 90% confidence-level bands. The big black dot is a long-term forecast by Agnus Maddison.[12]

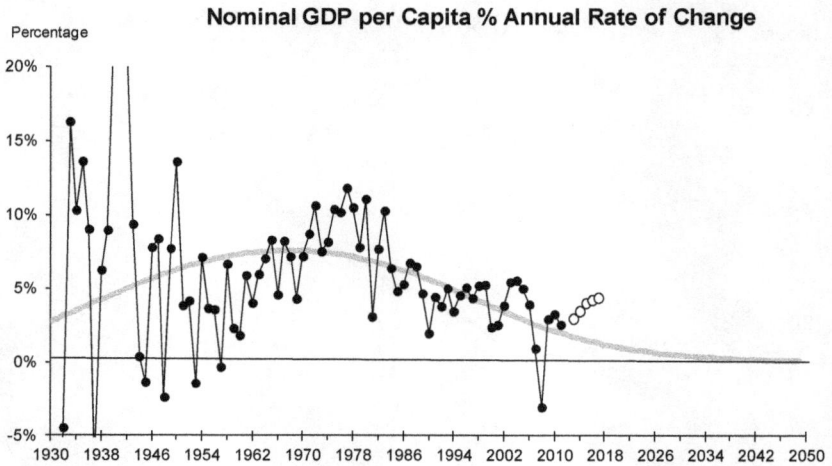

FIGURE 2.3 Annual percent rate of growth of nominal GDP per capita (black dots) and of the S-curve fit (thick gray line). The open circles are IMF forecasts.

In fact, the decline of nominal GDP's rate of growth in percentage terms began already in the late 1970s, as can be seen in Figure 2.3. The data fluctuate considerably particularly during early 20th century because of the small absolute values of GDP (which is in the denominator). Nevertheless, the thick gray line, derived from the S-curve in Figure 2.1, provides a good description for what happened during the last 80 years. The trend indicated by the IMF forecasts seems to be in sharp disagreement. Obviously, a forecast consisting of a constant annual percentage can be proven only wrong in the years to come. The end of US GDP growth seems to be scheduled for the mid 2021st century.

Inflation

Nominal GDP is related to real GDP via inflation. Therefore we can extract an overall trend for inflation from the two S-curves in Figures 2.1 and 2.2. This is done in Figure 2.4; the thick gray line is not a fit to the data here. It is calculated by dividing the gray S-curve in Figures 2.1 by the one in Figure 2.2, and seems to be a fair description of the trend of the consumer price index (CPI) over the past 80 years. It provides a long-term forecast for inflation—something of a Holy Grail quest for economists. It heralds deflationary times in the future, for which there is no indication in the evolution of the historical data. Is it believable?

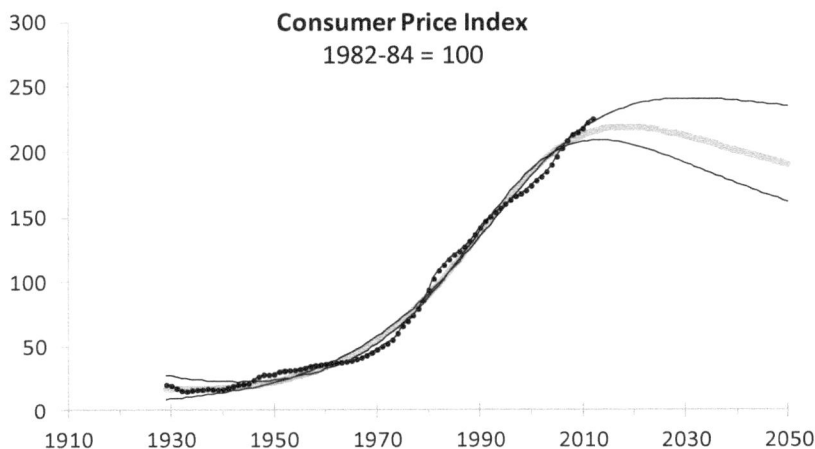

Consumer Price Index
1982-84 = 100

FIGURE 2.4 The black dots are the actual numbers for the consumer price index (1982-84 =100). The thick gray line indicating the inflation trend is not a fit to the data. It is calculated from the S-curves in Figures 2.1 and 2.2. The thin black lines delimit an uncertainty range resulting from the bands defined by the thin black lines in Figures 2.1 and 2.2.

The gray-line pattern of Figure 2.4, namely a long rise followed by flattening and a decline, depicts a rather typical behavior for inflation. I have seen it reproduced in many countries, and sometimes—for example the cases of Japan and Taiwan—the data also populated the declining phase of the pattern. Such a pattern can be understood as follows.

Inflation is harmful only when the CPI increases. Imagine a country where the CPI remains constant over time. Then nominal and real GDP will follow identical patterns except for a multiplicative constant. But if the CPI grows at the same time the real GDP grows, the nominal GDP will be made to grow faster. Therefore the nominal-GDP S-curve typically approaches a ceiling earlier than the real-GDP S-curve. And the ratio of the two S-curves (CPI trend) will follow the typical pattern depicted by the gray-line in Figure 2.4.

Socially speaking when GDP grows rapidly there is period of fat cows, growth and prosperity, people spend money because they can afford it, and prices go up. When GDP growth begins to saturate, there is period of skinny cows, people cannot afford high prices and inflation drops. That's why the CPI has a trajectory that goes up during the years of rapid GDP growth and then goes down when GDP growth saturates.

It makes sense that the inflation cycle goes over a peak and that its growth phase coincides with the rapid-growth phase of the GDP.

GDP Growth in Other Parts of the World

It is of interest to also examine the long-term prospects of GDP growth from a natural-growth point of view in other parts of the world. Figure 2.5 shows that the nominal GDP per capita in Japan has completed its S-curve twenty years ago! No uncertainties around this S-curve fit.

There has been no growth in Japan for twenty years and the traditional forecast from IMF for years 2013-2017—shown here with the open circles—does not contradict our description unless this rising trend continues well beyond 2017.

The evolution of GDP growth in Japan is a clear demonstration that GDP is a natural-growth process and will follow an S-curve to the end just like a species populations does in nature when it fills its ecological niche.

However in developing countries the story is different. Figures 2.6 and 2.7 show the nominal GDP per capita and S-curve fits for China and India respectively. The data come from EconStats.[13] By the end of 2012 both growth processes were in the very early stages of natural growth, a region where it is difficult to distinguish an S-curve from a simple

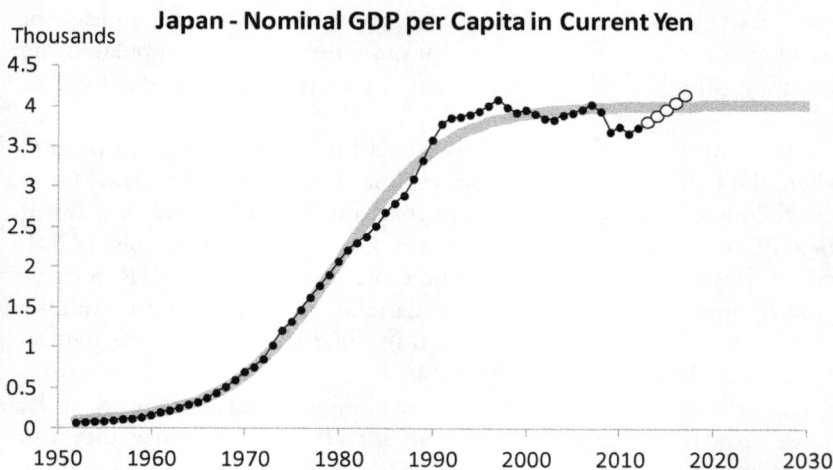

FIGURE 2.5 Japan nominal per capita GDP (black dots), IMF forecast (open circles) and S-curve fit (thick gray line).

Thousand Yuan
of 1990

China - Nominal GDP per Capita in Current Yuan

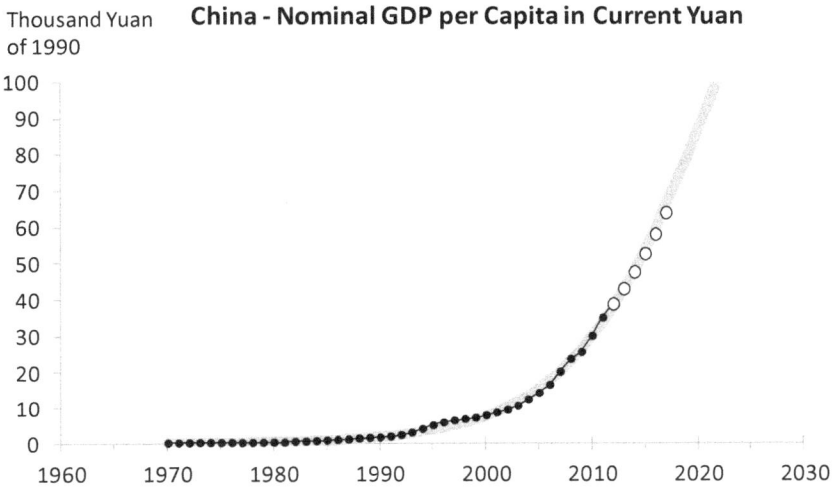

FIGURE 2.6 Nominal GDP per capita in China (black dots), IMF forecast (open circles) and S-curve fit (thick gray line.

Thousand ruppes
of 2004-2005

India - Nominal GDP per Capita in Current Rupees

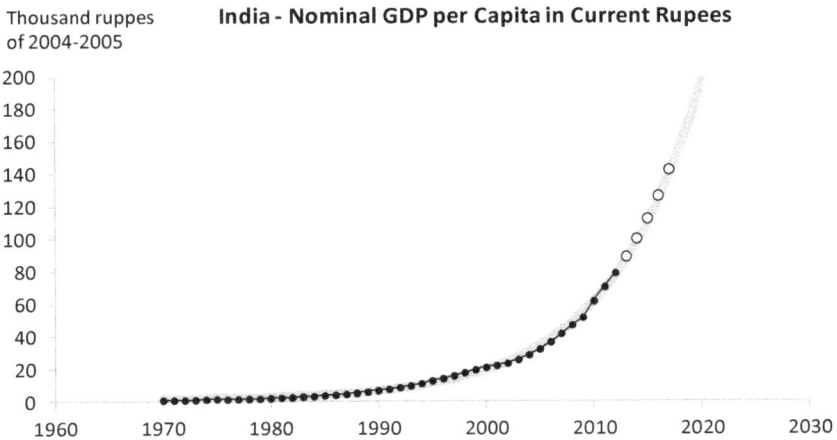

FIGURE 2.7 Nominal GDP per capita in India (black dots), IMF forecast (open circles) and S-curve fit (thick gray line).

exponential pattern. It is not realistic to try and establish uncertainties for the ceilings of the S-curves in Figures 2.6 and 2.7 because the patterns are still exponential for all practical purposes. Instead we can try to establish upper and lower limits for the ceilings of the S-curves these exponentials may eventually turn into. The discussion in Appendix D sets a lower limit as a factor of 5 on today's levels. Infant mortality and common sense can help us establish an upper limit. Infant mortality is usually taken between 5% and 10% of the final ceiling. A tree or a sunflower seedling of height less than 5% of its final size is vulnerable to herbivores or simply to be stepped on by a bigger animal. Assuming China's and India's natural-growth curves have advanced to at least beyond infant-mortality levels, their remaining growth potential should be at most about 10 times today's GDP levels.

The significant uncertainty range estimated on the ceiling of an eventual S-curve would have a rather small effect on the growth prospects during the next few decades. There is little doubt that China's remaining growth potential is a large factor over that of the US's and infinite over that of Japan's.

Discussion

One cannot assume that a country's economy, its wealth, its prosperity, or the productivity of its people will be growing indefinitely. These are natural-growth processes and they will eventually reach ceilings. The above analysis focuses on GDP as a measure of economic growth because it is the most frequently quoted metric and data are readily available worldwide. Analyzing other indexes of national progress, such as GPI (General Progress Indicator), would probably lead to similar conclusions. In Japan the nominal GDP per capita has already reached its ceiling in the early 1990s and the country has had rates of growth around null ever since. In the US the annual GDP increments have already entered diminishing trends for both nominal and real GDP per capita, more evidently for the former. The rate of growth for nominal GDP in percentage terms is expected to progressively diminish to 1.1% by 2020 and to 0.5% by 2030. All this should happen on the average, of course, as recessions and periods of growth come and go.

Considering that inflation is calculated from two natural-growth curves, it must also have a *natural* course to follow. In fact the CPI is forecasted to shortly enter a gentle downward trend, again on the average.

In view of an S-curve analysis, as a niche becomes full the rate of growth progressively drops to zero. This is in general the case with the evolution of GDP in industrialized countries. Barring catastrophes and disasters of unseen magnitude one should not expect renewed growth rates in the industrialized world. In contrast, developing countries like China and India, where the natural-growth process is still in its very early stages, should be expected to experience accelerating growth for decades.

An S-curve description may not be appropriate for the evolution of GDP in some situations. For example, one could have thought that undergoing such major "mutations" as Germany's reunification in 1991 and the European monetary union in 1992 would invalidate an S-curve description for the evolution of GDP in the countries involved. But apparently these mutations were not important enough. Most EU countries—including Germany—show only a minor glitch on the evolution of their GDP pattern around 1992; others like Belgium show no deviation whatsoever. Even Germany's reunification shows up as a minor deviation. In contrast, World War II breaks up the evolution of Germany's real-GDP-per-capita curve in order to place the country on an entirely different S-curve after World War II, see Figure 2.8.

Most countries in the European Union are experiencing saturation—i.e. they are approaching the ceiling of an S-curve—comparable to that of the US and Japan and this could be one explanation of the West's lingering economic malaise. For these countries to find themselves back on steeply rising growth patterns a very fundamental change must take place. Probably nothing lesser than the acquisition of new territory or the complete revamping of their economy—the way it happened in the

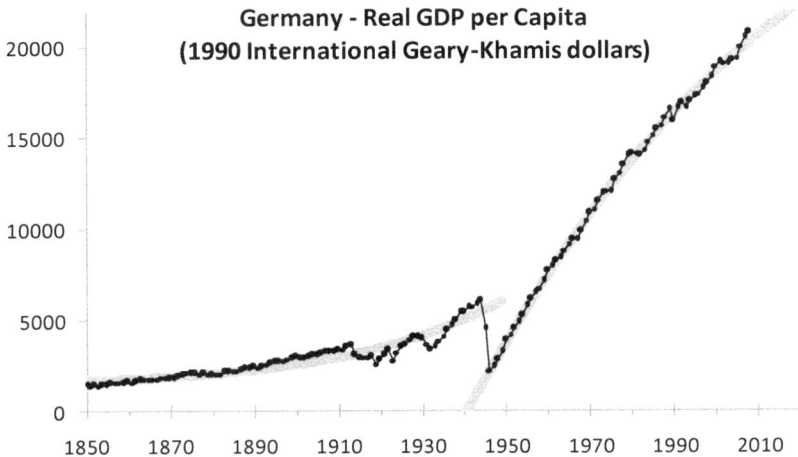

FIGURE 2.8 German real GDP per capita (black dots) and S-curve fits (thick gray lines). The data come from Agnus Maddison.[12]

US following the crash of 1929—would do the job. After all, as we can see in Figure 2.1, events as important as World War II, which had such a profound effect on the evolution of Germany's GDP, produced only an insignificant glitch on the smooth evolution of America's GDP.

THE COMING OF AGE OF THE UNITED STATES

• • •

"People at present think that five sons are not too many and each son has five sons also, and before the death of the grandfather there are already 25 descendants. Therefore people are more and wealth is less; they work hard and receive little."

Han Fei-Tzu 500 B.C.

• • •

As the number of developing countries that enter and grow in the world market increases the relative importance of developed countries—whose growth slows down anyway because their S-curves approach ceilings—becomes further diminished. We thus encounter a phenomenon of loss of market share not necessarily related to loss of competitiveness. This is clearly visible with the erosion of American dominance worldwide in a broad range of variables.

In what follows we will see a number of bell-shaped curves. They are the life cycles of S-curves fitted to the corresponding cumulative variable. But these bell-shaped curves are shares expressed in percentages, which makes difficult to understand the meaning of the corresponding S-curves and the underlying natural-growth-in-competition processes.

An S-curve implies competition for a limited resource. A competitor's market share cumulated over time will be limited because in a multi-competitor arena one competitor cannot be expected to hold on to his/her market share forever. So under the assumption that there will be an eventual decline to a competitor's performance, cumulative market share—expressed in some arbitrary units—becomes a variable amenable to an S-curve description.

In Figure 2.9 we see the evolution of the market share of the US in the world energy-consumption market. The gray life-cycle line describes the data well even if it has not been fitted to these data (an S-curve has been fitted to the cumulative market share). The lifecycle reaches a

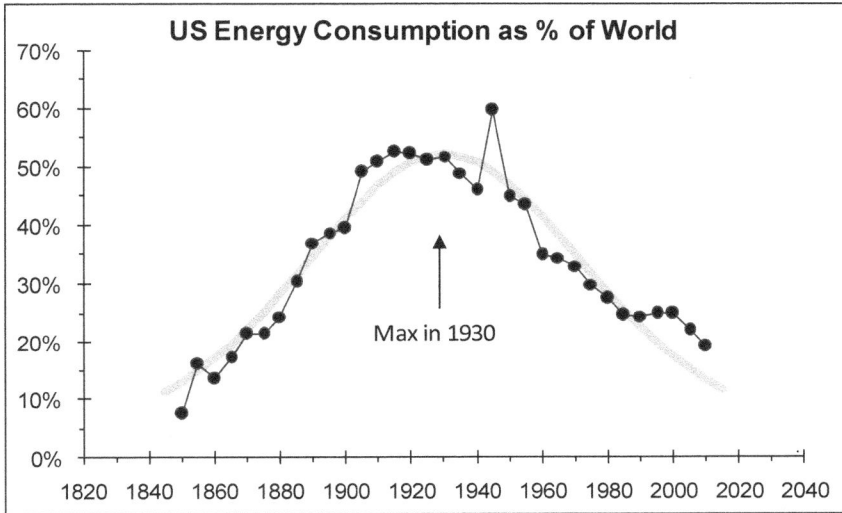

US Energy Consumption as % of World

FIGURE 2.9. The gray line is not a fit to the data. It is the life cycle of an S-curve that was fitted to the cumulative market share. The arrow shows the maximum of the gray bell-shaped curve.[14]

maximum in 1930 and then enters a smooth decline with exception an understandable short-term surge around WWII.

It is of interest that the apogee of US dominance in energy consumption worldwide coincides with the beginning of the Great Depression that was triggered by the stock-market collapse in New York City in late 1929. The Great Depression was felt around the world and many countries suffered more severely than the US, and yet the American share of energy consumption embarked on a declining trend that it has been following ever since.

In Figure 2.10 we see the evolution of the share of the US in the world population. The US population has been growing not only because the fertility of American women has generally been above the 2.1 children necessary to maintain the population size, but also because of immigration. Despite the fact that immigration played an important role in the 19th century and early 20th, the annual increase of American population due to immigration has been steadily growing since the late 1940s with a marked peak around 1991. And yet, as percentage of world population the US reached a maximum in 1942 and has been declining ever since. Fertility plus immigration have not kept up with the rate of growth of the world's population.

In view of a generally small share of the world population—less than 6%—the energy consumption share we saw in Figure 2.18 becomes rather impressive; it says that the average American spends about ten times more energy than the average world citizen.

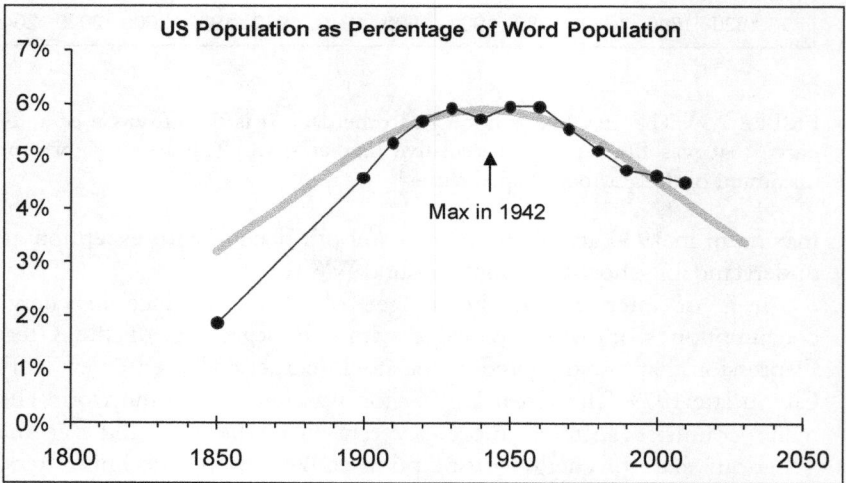

FIGURE 2.10 The gray line is not a fit to the data. It is the life cycle of an S-curve that was fitted to the cumulative share. The arrow shows the maximum of the gray bell-shaped curve.[15]

Figure 2.11 shows the evolution of the share of Olympic medals won by Americans. Here we have excluded the years when the games were held on American soil because of the well-known phenomenon of the home country winning disproportionally many medals. We see again that the American share increased from the beginning of the games in 1996 only to decline during the second half of the 20th century. Even though the data show a peak in 1924, the theoretical maximum of the bell-shaped curve is in 1944 when there were no games because of World War II. During the second half of the 20th century the American share steadily decreased along the life-cycle trajectory, quite similar to what the population share did in Figure 2.19. This is not surprising because a larger population offers a greater base from which exceptional athletes can emerge. However, the fact that the Olympic-medals share is four times bigger than the population share makes the US's athletic performance stand out well above the average country around the world.

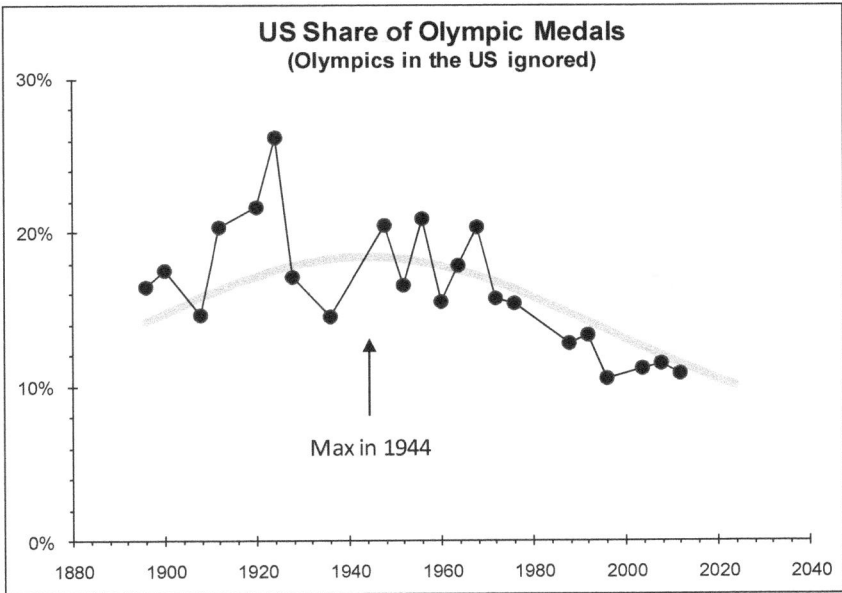

FIGURE 2.11 The gray line is not a fit to the data. It is the life cycle of an S-curve that was fitted to the cumulative market share. The arrow shows the maximum of the gray bell-shaped curve.[16]

A country's role in the world's economy is perhaps best illustrated by the country's share of the world's GDP. Figure 2.12 shows the US GDP claimed a progressively greater share of the world's GDP up to 1960 and has been declining ever since. This reflects the years of economic boom the US experienced in the 1950s and the 1960s.

One may have thought that GDP growth should go hand in hand with energy consumption and yet we see a thirty year delay between the peaks in Figure 2.9 and Figure 2.12. It would seem that it takes about one generation between the time a nation works hard and the time its economy booms. This should come as news to politicians who aim to rapidly grow the economy with stimulus packages.

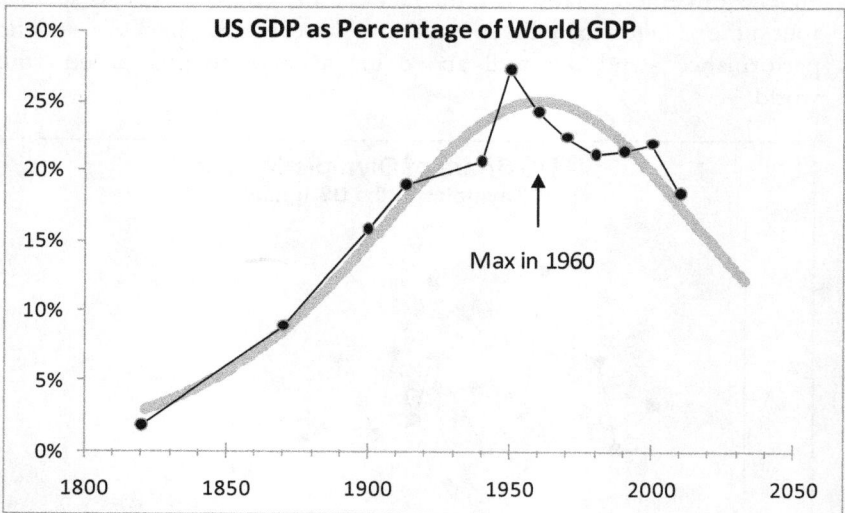

FIGURE 2.12 The gray line is not a fit to the data. It is the life cycle of an S-curve that was fitted to the cumulative market share. The arrow shows the maximum of the gray bell-shaped curve.[17]

A country's ability to produce industrial innovation has often been cited as a reason for successful economic development. And yet, in Figure 2.13 we see that US patent applications as percentage of world patent applications peaks in 1995, i.e. about a generation later from when the US GDP peaked.

It would seem that it is prosperity which provides a fertile ground for ingenuity and technological breakthroughs, and not the other way around. Of course, the fact that prosperity preceded patents does not constitute proof that wellbeing triggers inventiveness but it disproves the opposite, namely the argument that enhanced patent production triggers economic growth. In this content patents take on the appearance of a pastime in an affluent society!

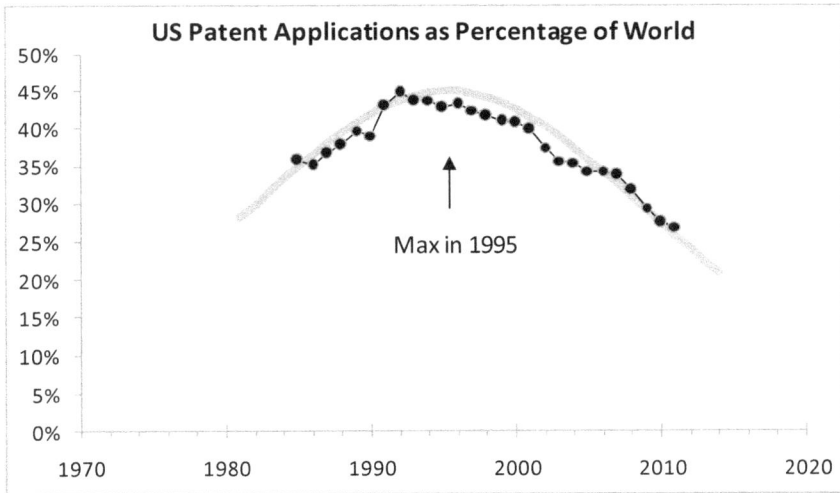

US Patent Applications as Percentage of World

Max in 1995

FIGURE 2.13 The gray line is not a fit to the data. It is the life cycle of an S-curve that was fitted to the cumulative market share. The arrow shows the maximum of the gray bell-shaped curve.[18]

Summarizing the results presented in this chapter we may say that the evolution of the United States has in many respects gone over the successive phases of rapid growth, peak, and decline. In the case of GDP most people have not yet become aware of this trend's systematic decline. On the world scene America's share rose, reached a peak, and declined in importance first in energy consumption (1930), then in population (1942), then in Olympic medals (1944), then in GDP (1960), and finally in patent applications (1995). This succession may be suggestive of causal relationships, namely that energy consumption may stimulate a population increase, which in turn may reflect in winning more Olympic medals; or that GDP growth triggers more patent applications. But preceding something does not necessarily mean causing it. It does, however preclude that what followed caused what preceded, that is, we can say with certainty that population growth did not cause energy consumption to increase, and that GDP growth and affluence did not contribute to winning more Olympic medals.

It is of interest that intellectual performance declines last. Besides patent applications that peaked in 1995, the American share of Nobel laureates has not begun declining yet, as we will see in Figure 3.7 in next chapter. Like a living organism that is aging, the "physical" aspect of the United States seems to be wearing down before the "intellectual" one.

3

Substitutions

A hard fact of life: young will replace old.

• • •

In its simplest form natural growth in competition is a process in which one or more "species" strive to increase their numbers in a "niche" of finite resources. Depending on whether or not the "species" is successful over time, its population will trace an ascending or a descending S-curve. In a niche filled to capacity, one species population can increase only to the extent that another one decreases. Thus occurs a process of substitution, and to the extent that the conditions of competition are natural, the transition from occupancy by the old to occupancy by the new should follow the familiar S-shaped pattern of a natural growth process.

The first connection between competitive substitutions and S-curves was done by J. C. Fischer and R. H. Pry in a celebrated article published in 1971. It became a classic in studies of the diffusion of technological change. They wrote as follows:

> If one admits that man has few broad basic needs to be satisfied—food, clothing, shelter, transportation, communication, education, and the like—then it follows that technological evolution consists mainly of substituting a new form of satisfaction for the old one. Thus as technology advances, we may successively substitute coal for wood, hydrocarbons for coal, and nuclear fuel for fossil fuel in the production of energy. In war we may substitute guns for

arrows, or tanks for horses. Even in a much more narrow and confined framework, substitutions are constantly encountered. For example, we substitute water-based paints for oil-based paints, detergents for soap, and plastic floors for wood floors in houses.[1]

They went on to explain that depending on the timeframe of the substitution the process may seem evolutionary or revolutionary. However, regardless of the pace of change, the end result is to perform an existing function or satisfy an ongoing need differently. The function or need rarely undergoes radical change.

A similar process occurs in nature, with competition embedded in the fact that the new way is vying with the old way to satisfy the same need. And the eventual winner will be, in the Darwinian formulation, the one better fit for survival. This unpalatable conclusion may seem unfair to the aging, wise, and experienced members of human society, but in practice it is not always easy for the young to take over from the old. Experience and wisdom do provide a competitive edge, and the substitution process proceeds only to the extent that youth fortifies itself progressively with knowledge and understanding. It is the combination of the required proportions of energy, fitness, experience, and wisdom that will determine the rate of competitive substitutions. If one's overall rating is low due to a lack in one area, someone else with a better score will gain ground, but only proportionally to his or her competitive advantage.

The opposite substitution, cases in which the old replace the young, is possible but rare. It is sometimes encountered in crisis situations where age and experience are more important to survival than youth and energy. But independently of who is substituting for whom, and in spite of the harshness ingrained in competitive substitutions, one can say that the process deserves to be called natural.

ONE-TO-ONE SUBSTITUTIONS

A species population growing into an ecological niche that is already filled to capacity will progressively replace the incumbent species. A classical example was the replacement of horses by cars as means of personal transportation at the beginning of the twentieth century.

When automobiles were first introduced, they offered an appealing alternative to traveling on horseback. The speed and cost, however, were not terribly different. In *Megamistakes: Forecasting and the Myth of Rapid Technological Change*, Steven Schnaars considers a cost-benefit analysis as one of the three ways to avoid wrong forecasts.[2] In this case such an analysis would have predicted a poor future for cars.

Ironically, one of the early advantages of the automobile was its non-polluting operation. Cleaning the streets from horse excrements in large metropolises was becoming an increasingly laborious task, and forecasts for the transport needs of the future rendered the situation alarming. For this and mainly for other more deeply-rooted reasons, cars began replacing horses rapidly and diffused in society as a popular means of transportation.

We can look at this substitution process in detail by focusing on the phasing-in period of the automobile 1900-1930. Let us look at only the relative amounts, that is, the percentage of cars and horses of the total number of transport "units" (horses plus cars.) Before 1900, horses filled 100% of the personal-transport niche. As the percentage of cars grew, the share of horses declined, because the sum had to equal 100%. The data in Figure 3.1 show only non-farming horses and mules.

These trajectories are seen to follow complementary S-curves. Around 1916 there are an equal number of horses and cars on the streets, and by 1925 the substitution is more than 90% completed. It is

The Substitution of Cars for Horses in Personal Transportation

As a percentage off
all "vehicles

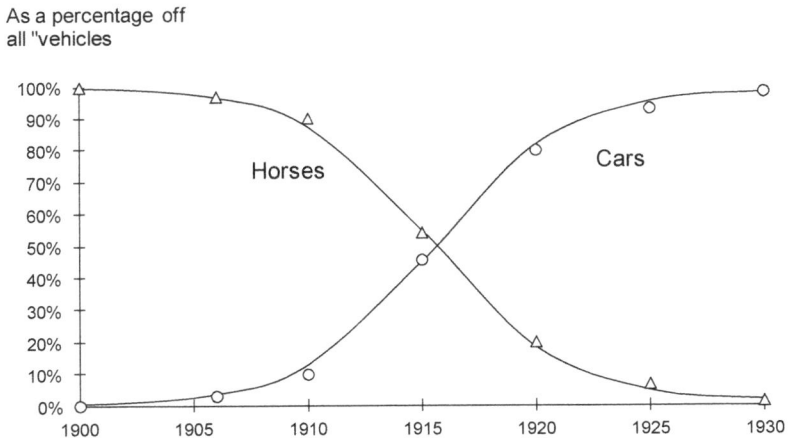

FIGURE 3.1   The data points represent percentages of the total number of transportation units, namely, cars plus non-farming horses and mules. The S-curves are fitted to the data points. The sum of respective ascending and descending percentages equals 100% at any given time.[3]

interesting to notice that the fitted S-curve does not quite reach the ceiling of 100% for cars after 1930. This may be related to the fact that a certain number of horses were not replaced by cars. They are probably the horses found today in horseback riding, horse racing, and in tourist attractions.

The obsolescence of horse as a means of transportation illustrates the inevitable takeover by newcomers possessing competitive advantages. There are always two complementary trajectories in one-to-one substitutions, one for the loser and one for the winner. They indicate the shares, the relative positions, of the contenders. Because the ceiling is by definition 100%, the determination of these curves becomes easier and more reliable than usual S-curves. The niche size for shares is by definition 100% and it is not a function of time.

By looking at shares we focus on the competition and we evidence an advantage whose origin is deeply rooted, as if it were genetic in nature. This description thus becomes free of external influences: the state of the economy, politics, earthquakes, and seasonal effects such as vacations and holidays. In the case of the cars-for-horses substitution, World War I had no impact; the trajectories in Figure 3.1 continued their smooth evolution undisturbed.

Another advantage of focusing on relative positions in substitution processes is that in a fast-growing market the natural character of the process (the shape S) may be hidden under the absolute numbers that are increasing for both competitors. But the fact that one competitor is growing faster implies that the other competitor is phasing out. During the first decade of the twentieth century the number of horses continued to increase as it had in the past. The number of cars increased even more rapidly, however. The substitution graph of Figure 3.1 reveals an indisputable decline for the horses' share during the decade in question.

The S-curve equation used in the one-to-one natural substitutions is such that if one divides the number of the "new" by the number of the "old" at any given time and plots this ratio on a logarithmic vertical scale, one gets a straight line. It is a mathematical transformation. The interest in doing this is to easily detect "naturalness" in substitution processes. The S-curve pattern will be made to appear as a straight line when viewed through this "eyepiece." In theory even only three data points that fall on a straight line indicate the existence of an S-curve, albeit with poor confidence (the more points we have, the higher our confidence). This eliminates the need for computers and sophisticated fitting procedures in order to "see" S-curves in substitution processes. All we need to do is look for a straight-line pattern in the data, the existence of which will be proof that indeed we are dealing with natural growth in competition.

Transforming an S-curve into a Straight Line

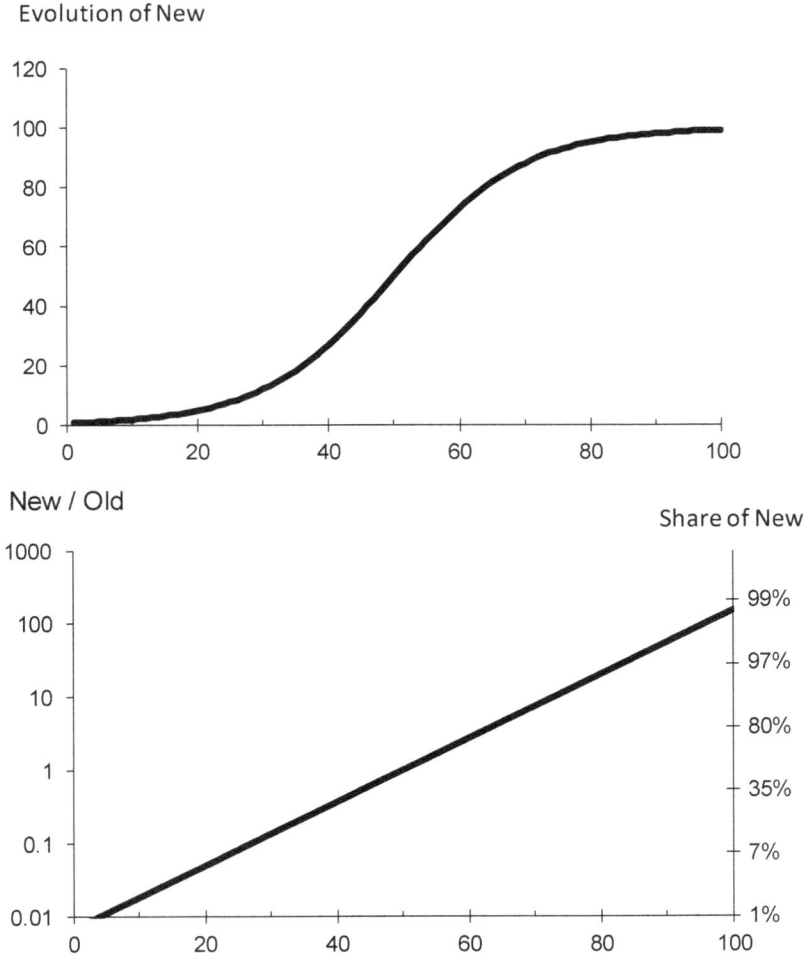

Evolution of New

New / Old

Share of New

FIGURE 3.2 The S-curve at the top is transformed into a straight line at the bottom by plotting the ratio "new" to "old" on a logarithmic scale at the left while the share of "new" is designated by the nonlinear "logistic" scale at the right. This scale is nonlinear with 100% infinitely far upward and 0% infinitely far downward.

Looking at natural substitution processes in this way reveals straight lines. The straighter the lines, the more natural the processes. The longer the lines, the more confidence one can have in extrapolating them further. If the ratio of new to old has not yet reached 0.1, it may be too early to talk of a substitution; a natural substitution becomes "respectable" only after it has demonstrated that it has survived "infant mortality" as explained in Chapter 2.

Another reason for which an *a priori* natural substitution may not display the expected trajectory is that the variable is not properly defined. For example the substitution of detergent for soap discussed in the article by Fisher and Pry would not have followed a straight line had they taken all soap. They carefully considered only laundry soap, leaving cosmetic soap aside. Complications can also arise in substitutions that look "unnatural" in cases of a niche-within-a-niche, or a niche-beyond-a-niche. In both cases two different S-curves must be considered for the appropriate time periods. In the straight-line representation such cases show a broken line made out of two straight sections.

When close scrutiny does not eliminate irregularities, it means that there is something unnatural after all. A substitution may show local deviations from a straight-line pattern, which can be due to exceptional temporary phenomena. But such anomalies are soon reabsorbed, and the process regains a more natural course.

Among the many examples of one-to-one substitutions quantitatively discussed in *Predictions* are the substitutions of synthetic for natural rubber, synthetic for natural fibers, and margarine for butter in the United States. All three depict the same slope (the three straight lines are parallel). This does not mean that all substitutions always follow this rate, but it does point to a business-as-usual aspect of society, and not only American society. The detergent-for-soap substitution mentioned earlier took place at the same rate in Japan as it had done in the United States ten years earlier.[4]

However, the replacement of natural by synthetic rubber shows a large "anomaly" during the war years. In the 1930s synthetic rubber was slowly making an appearance in the American market as a more expensive alternative of lesser quality than natural rubber, which was imported in large quantities from abroad. During the early stages of World War II, the imports of natural rubber were severely diminished and, at the same time the demand for rubber was significantly increased. As a consequence much effort was put in improving the quality and reducing the production costs of synthetic rubber.

Synthetic replaced natural rubber at an accelerated rate during the war years as can be evidenced in Figure 3.3. But as soon as the war ended, foreign sources of natural rubber became available again, and the substitution rate dropped despite the technological progress that had been made. From then on the substitution process continued at a rate similar to other replacements, such as margarine for butter and synthetic for natural fibers to reach again war levels of production only twenty years later. The deviation, caused by the necessities of the war, disappeared leaving no trace when life got back to normal.

Wars May Interfere with Natural Substitutions

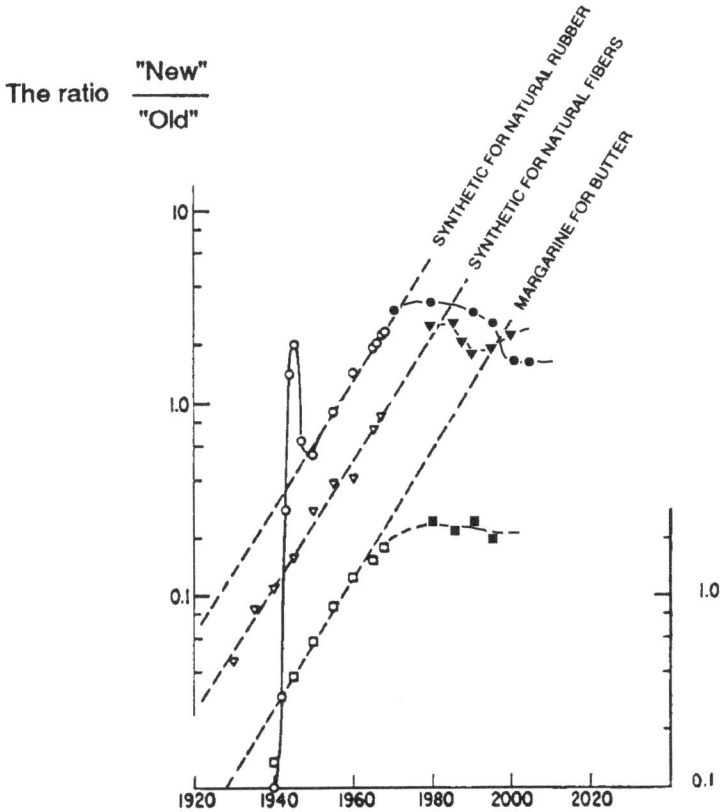

FIGURE 3.3 Three replacements in the United States published by J. C. Fisher and R. H. Pry in 1971. The data show ratios of amounts consumed. The logarithmic scale brings out the straight-line character of the substitutions. The only significant deviation from a natural path is observed for the production of synthetic rubber during World War II. The solid-black points show recent data.[5]

The solid-black points in Figure 3.3 show that none of the three substitution processes finally reached 100% completion. They all stopped at around 90%. Contrary to what may have been expected from the law of natural competition one-to-one substitutions do not always proceed to completion. A "locked" portion of the market share may resist substitution. In our three examples this locked portion was about 10% of the market; it could be larger. Competitive substitutions are natural growth processes and do follow S-curves, but the newcomer rarely replaces the incumbent completely.

SUCCESSIVE SUBSTITUTIONS

The introduction of a gifted new competitor in an already occupied niche results in a progressive displacement of the older tenant, and the dominant role eventually passes from the old to the new. As the new gets older, however, it cedes leadership in its turn to a more recent contender, and substitutions may thus cascade. The following examples had been analyzed in *Predictions* and are updated now.

Diseases

We can think of diseases as species of microorganisms, the populations of which compete for growth and survival. But with the overall number of potential victims being limited, the struggle will cause some diseases to grow and others to decline, a situation similar to different species populations growing to fill the same ecological niche. The extent of the growth of a disease can be quantified by the number of victims it claims. A relative rating is obtained if we express this number as a percentage of all deaths. The percentage of victims claimed by the disease best fitted for survival increases every year, while the unfit disease claims a declining share. Young diseases are on the rise while old ones are phasing out.[6]

In *Predictions* I charted this phenomenon and obtained a simplified but quantitative picture in which diseases were grouped into three broad categories according to prevalence and nature, see Figure 3.4. One group contained all cardiovascular ailments, today's number-one killer. A second group was called neoplastic and included all types of cancer. The third group comprised all old, generally phasing out, diseases. Hepatitis and diabetes were not included because their share was rather flat over time (no competitive substitutions) and below the 1 percent level (not visible in the usual picture). Cirrhosis of the liver also represented too small a percentage, but being of an intriguing cyclical nature, it is reported separately later in Figure 8.4 in Appendix A.

In Figure 3.4, the vertical scale is again *logistic* so that S-curves become straight lines. From the turn of the century, cardiovascular ailments have been claiming a progressively increasing share, making them the dominant cause of death by 1925. Their share reached a peak in the 1960s with more than 70 percent of all deaths and then started declining in favor of cancer, which had also been steadily growing at a comparable rate—parallel line—but at a lower level. All other causes of death had a declining share from the beginning of the century. Today cardiovascular ailments may claim many more victims than cancer, but their share is steadily decreasing in favor of the latter. Projections of the natural trajectories made with data up to 1988 indicated that cancer and cardio-vascular diseases would split the total death toll fifty-fifty around 2020. The small circles generally confirm the forecasts made twenty years ago. The irregularities in the declining trajectory of Other are at the few percent level and therefore are not significant (the vertical scale is very non-linear—logistic).

The smooth lines in Figure 3.4 represent the description given by the substitution model as generalized by Nakicenovic to include more than two competitors.[7] It says that at any given time the shares of all competitors but one will follow straight-line trajectories. The singled out competitor, called the saturating one, will have its share determined as the remainder from 100 percent after all other shares have been subtracted. The saturating competitor is normally the oldest among the ones that were still growing. Usually it is also the one with the largest share, has practically reached maximum penetration, and is expected to start declining soon. The trajectory of a saturating share is curved. It traces the transition from the end of the growing phase, where the entity was substituting for its predecessor, to the beginning of the declining phase, where it will itself start being replaced by the one next in line. Every competitor goes through the saturating phase in chronological order.

The saturation phase can be seen as the culmination and, at the same time, as the beginning of the end. For a product it often corresponds to the maturity phase of the life cycle. It is the time when you face competition directly. You have done well, and everyone is trying to chip away at your gains. Your share is calculated as what remains after subtracting from 100 percent the trajectories of all the others. You are at the top, but you are alone; the model requires one and only one in the saturating phase at a time. This condition is necessary in order to produce

Competition Between Diseases

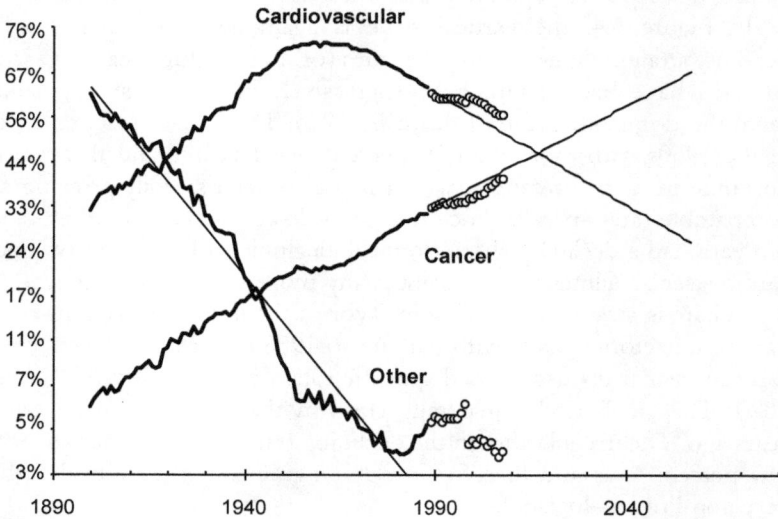

FIGURE 3.4 This figure splits the total number of deaths in the United States into major disease categories. Once again, the vertical scale is *logistic*. We see only the projections of the straight lines fitted on the data. The small circles generally confirm the predictions made with data up to 1988. The small upward trend of Other in recent years seems important because of the nonlinear scale, but it only amounts to about 1% gain.[8]

a workable model, but it also matches well what happens in a multi-competitor arena—typically, the Olympic Games. There is always one distinguished as first runner even if required to use precision watches with time resolutions of 1/100 of a second or better.

Means of Transportation

Another successive substitution concerns means of transportation. It was shown in *Predictions* that car populations were to saturated their niche in society for most European countries, Japan, and the United States by the end of the twentieth century. We are at the end of the era in which people were preoccupied by the automobile. After all, air travel has been gaining importance relative to road travel ever since 1960, when the latter enjoyed the lion's share of the transportation market.

In 1960 the automobile was at its zenith. In the division of US intercity passenger traffic among trains, cars, and airplanes, shown in Figure 3.5, the

Dressing Up as the Devil

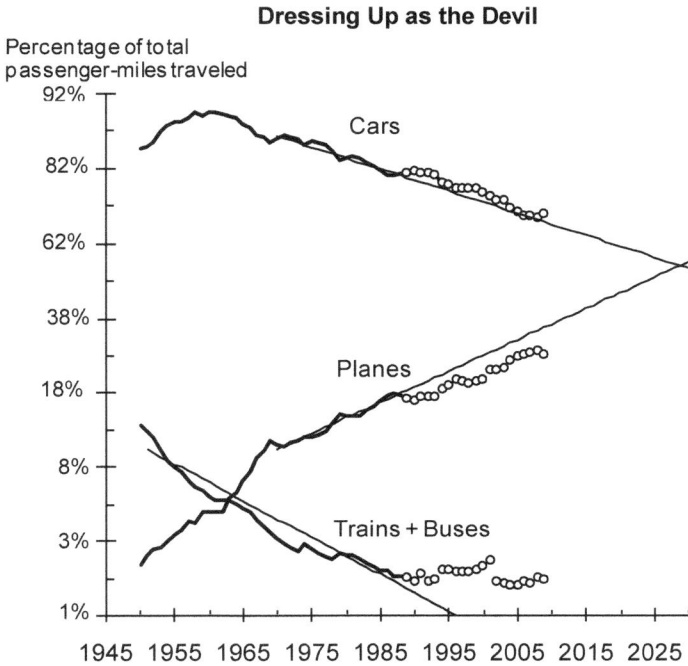

FIGURE 3.5 On the top we see the division among means of transportation competing for intercity passenger traffic in the US. The vertical scale is logistic. The straight lines are extrapolations of fits on the data. The small circles show what happened since 1988. On the bottom, a 1959 Cadillac Cyclone is representative of car "behavior" at the moment when car dominance in this market started being challenged by airplanes. Adapted from work carried out at the IIASA.

percentage of traffic attributed to trains (buses are included in this category) has been systematically declining, while that of airplanes has been rising. The share of cars rose until 1960, reaching close to 90 percent, but then it began to decline. Cesare Marchetti believes that the automobile, in spite of its dominant position at the time, "felt" the rising threat of airplanes. He whimsically adds that at the moment when the automobile's market share started declining, cars were "masquerading themselves as airplanes with Mach 0.8 aerodynamics, ailerons, tails, and 'cockpit' instrument panels. The competitor is the devil—in this case the airplane—and as peasants still do in the Austrian Alps, to scare the devil, one has to dress like the devil."[9]

The small circles in Figure 3.5 show a small deviation from the predicted trajectories in recent years. This deviation, as we will see in the next section, has to do with the fact that traveling by air grew less rapidly than expected probably because aviation technology has not yet adopted a more hydrogen-rich fuel (e.g. natural gas or hydrogen). The takeover point, that is, the time when more passenger-miles will be covered in the air than on the ground must consequently be pushed back by ten or twenty years.

Air travel has undoubtedly a significant remaining growth potential. A graph of the ton-kilometers carried every year worldwide shows that air traffic is heading for a point of saturation toward the middle of the twenty-first century. The S-curve that best fits the data has a ceiling, which at the time of *Predictions* had been estimated at 580 billion ton-kilometers per year. But in Figure 3.6 we see that this ceiling had been significantly underestimated. Nevertheless, for fourteen years the data came in excellent agreement with the predicted trajectory. Only the last five data points indicate a clear deviation from the predicted trend. Conceivably air traffic underwent some "mutation" around 2003—experts should be interrogated about this—otherwise the deviation must be attributed to the known bias of S-curve forecasts for low ceilings. In any case, a forecast that proves accurate for fourteen years can be considered a success; like all forecasts S-curves must be updated after ten years or so.

In addition the agreement between the data and the S-curve description must be considered remarkable, given that there have been at least three sharp fuel-price increases, in 1974, in 1981, and in 2008. One may have expected that the diffusion of air traffic would be impacted by the price of fuel on which it depends so directly. Not at all! The system seems to compensate internally for such events so as not to deviate from its natural course.

The Growth of Air traffic Will Continue Well into the 21st Century

Billions of ton-kilometers
performed yearly

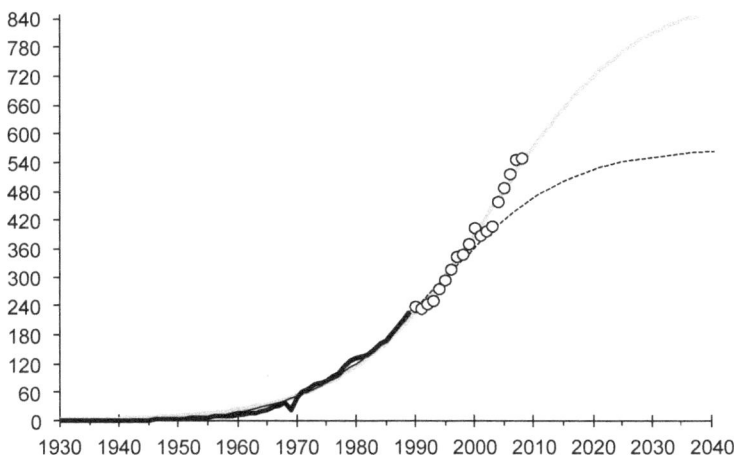

FIGURE 3.6 World air traffic data (thick line) and fit (light-gray line.) The intermittent line is the forecast that had been made with data up to 1989. The open circles are recent data not participating in the dotted-line S-curve fit.[10]

Nobel Prize Awards

The phenomenon of natural substitution can also be seen in operation with a more noble form of competition: Nobel Prize awards. The total number of Nobel awards per year can be thought of as a market, a pie to be distributed among the candidates. If we group together individuals with common characteristics or affiliations, we obtain regions or countries that become the contenders competing for prizes. Many countries can only boast a few award winners each, so grouping countries together becomes essential if one wants to look for trends. The most reasonable grouping comprises three regions. The United States is one such region. Another is "classical" Europe, consisting of France, Germany, Great Britain, the USSR, Italy, the Scandinavian countries, Belgium, Holland, Austria, Switzerland, Spain, and Ireland. The third group is what remains and may be called the Other World. It includes many Third World countries and many developing countries, but also Japan, Australia, and Canada.

This classification must be characterized as natural because when we graph the shares of the three groups as a function of time, we see straight lines emerging (in a graph with a logistic vertical scale) for the various substitutions; and there are three such substitutions, see Figure 3.7. The share of classical Europe accounting for 100 percent at the beginning of the twentieth century decreased along a straight line during the first four decades of the 20th century. The United States picked up the European losses, and its share rose along an almost complementary line. After World War II the American share stopped growing, the European share shifted to a less dramatic decline, and the Other world became the rising new contender.

Figure 3.7 was published in *Predictions* with data up to the end of 1988. Within that dataset the competitive substitutions had corroborated

Competition for Nobel Prizes

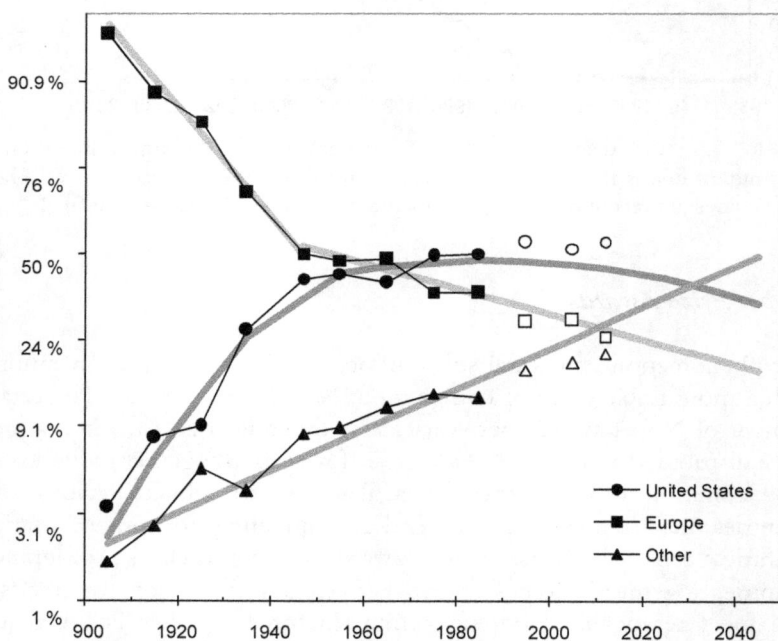

FIGURE 3.7 We see here the split of Nobel prize awards among three major regions in the usual representation. Complementary straight lines are evidence for three general substitutions: Europeans yielding first to Americans, then to Other, and finally Americans also declining in favor of Other. The last three data points on each variable indicate what happened since 1988 when the trends (gray lines) were first established.[11]

an age advantage. The American award winners had been gaining on the Europeans up to 1940, with an average age of 53.5 versus 55 years. Similarly, in the period 1975-1987 Other world winners gained on the Americans, with an average age of 58.7 versus 61 years, while European winners continued losing with an average of 63.6. The argument of age became even more relevant when we consider that the actual work was often done many years before receiving the award. The recipient's age at award time was chosen simply because it was more readily available and because in many cases it is better defined than the date when the actual work was done. The evidence for the age advantage is correct in any case as long as there is no bias among the three regional groups with respect to this delay. Youth, once again, seems to have an advantage over old age.

The last three points of recent data on each line generally confirm the trends first established in 1988. Any deviation the reader may want to see would be in the same direction as with Figure 1.4, namely highlighting an enhanced staying power for the US share on the expense of Other rather than Europeans.

FISHING AND PRYING

When Fisher and Pry put forth their model for competitive substitutions, they were mainly concerned with the diffusion of new technologies. Similar applications of the logistic function have been employed in the past by epidemiologists in order to describe the spreading of epidemic diseases. It is evident that the rate of new victims during an epidemic outbreak is proportional to both the number of people infected and the number remaining healthy, which spells out the same law as the one describing natural growth in competition. The diffusion of an epidemic is a substitution process in which a healthy population is progressively replaced by a sick one. In all three processes, diffusion, substitution, and competitive growth, the entities under consideration obey the same law.

In their paper Fisher and Pry looked at the rate of substitution for a variety of applications and found that the speed at which a substitution takes place is not simply related to the improvements in technology or manufacturing or marketing or distribution or any other single factor. It is rather a measure of how much the new is better than the old in all these factors. When a substitution begins, the new product, process, or service struggles hard to improve and demonstrate its advantages over the old one. As the newcomer finds recognition by achieving a small percentage of the market, the threatened element redoubles its efforts to maintain or improve its position. Thus, the pace of innovation may increase significantly during the course of a substitution struggle. The curvature of the bends and the steepness of the S-curve traced out by the ratio of total-market shares, however—the slope of the straight line in the logarithmic graph—does not change throughout the substitution. The rate reflected in this slope seems to be determined by the combination of economic forces stemming from the superiority of the newcomer.

Fisher and Pry suggested that this model could prove useful to investigations in the many aspects of technological change and innovation in our society. In fact, its use has spread far beyond that application. I doubt they realized back in 1970 how wide a range of applications their model would diffuse in during the decades that followed, and I agree with Marchetti when he says that they provided us with a tool for "Fishing and Prying" into the mechanisms of social living.

4

Where Has the Energy Picture Gone Wrong?

Under pressure society places hopes in forlorn directions such as bio-fuels, windmill farms, hybrid cars, and photovoltaic rooftops.

• • •

There is hardening evidence against such popular beliefs as the linking of oil prices to scarcity and the likening of renewable energies to panacea. There is also mounting pressure to urgently find the right direction for a concerted worldwide effort to meet energy needs. In what follows, a growth-in-competition approach involving S-curves produces far-reaching insights.[1]

US CRUDE OIL

We can look at the surfacing of oil as if it was a "population" growing to fill (or empty)—a "niche." The niche may be the amount of oil Mother Earth has in store underground for us. Alternatively, the niche may simply be the amount of oil for which we have a well-defined need.

Oil began its commercial production in 1859, but significant amounts of oil were extracted only from the early twentieth century onward. From the beginning, oil extraction triggered exploration for more oil reserves. But because exploration was expensive it was pursued only whenever it became necessary. Figure 4.8 shows both production and discovery of reserves for the United States. The historical data represent cumulative production and cumulative discovery of reserves.

Oil production gives rise to a smoother curve than discovery, which features random fluctuations due to the inherent randomness associated with a search. Both sets of data are amenable to good fits by S-curves (depicted by thick lines).

The two curves are remarkably parallel, with a separation of about ten years rather constant over most of the range. Such a rigid correlation between production and discovery over almost a whole century is proof of an underlying regulatory mechanism. The fact that oil production preceded oil discovery suggests that it may be production that triggers discovery! But chronological succession is not necessarily proof of causality. It does disprove the opposite however, namely that oil discovery gives rise to production. In any case, the regular succession of the two curves implies that we discover oil ten years before we consume it, not earlier or later.

This equilibrium has not resulted from some conscious decision. On the contrary, experts in oil have often forecasted imminent doom, with oil shortages and even depletion in a few years. In contrast, Figure 4.8 seems to indicate that the more you milk the reserves, the more reserves will be made available to you.

Oil Discovery and Production Go Hand in Hand

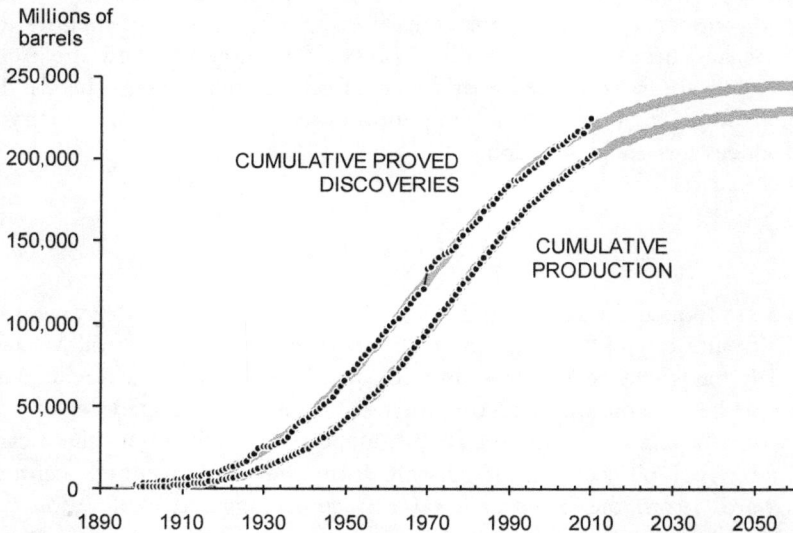

FIGURE 4.8 Yearly data and S-curve fits (thick grey lines) for oil discovery and production in the United States. The agreement between production data and the corresponding S-curve is impressive.[2]

The projections of the S-curves yield rather reliable forecasts given how closely and how extensively the two growth processes have followed the natural-growth pattern.[*] As we move into the future, the time difference between oil discovered and oil produced will progressively increase from today's ten years. Deep into the 21st century there will remain a permanent excess of proven oil reserves of about 15,000 million barrels that will never become object of production.

More extensive monthly data, shown in Figure 4.9, quantify the phasing out of oil production in the US. Today's level of production is less than half what it was in the mid 1970s. Interestingly, from the three oil shocks—in 1974, in 1981, and in 2008—only the last one seems to have left a mark on the evolution of the oil-production trend; an upward turnaround that nevertheless must be considered a short-live deviation to soon subside permitting the naturally declining trend to continue. Abandoning oil is not related to supply shortages. Other more competitive—better fit for survival—energy types are taking over.

It is noteworthy that when the decline of oil production in the US was predicted in *Predictions* with data up to 1982, it seemed hard to believe.

US Crude Oil Production (Monthly Rate)

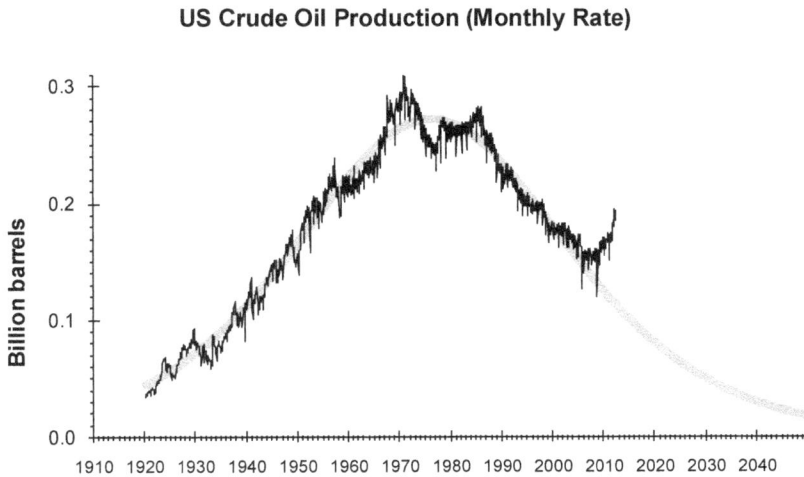

FIGURE 4.9 Monthly data. The gray life-cycle curve corresponds to the S-curve of Figure 4.8.[3]

---

[*] It must be noted, however, that the ceilings of these two S-curves with data up to 1982 was underestimated by about 10% in *Predictions*.

THE WORLD ENERGY PICTURE

On a worldwide scale the generalized substitution model gives another mountain like landscape when it comes to describing the primary-energy mix in energy consumption, see Figure 4.10. During the last one hundred and fifty years, wood, coal, natural gas, and nuclear energy have been the main protagonists in supplying the world with energy. More than one energy source was present at any time, but the leading role passed from one to the other. Wind power, waterpower, and other renewable sources have been ignored because they command too small a market share.

Figure 4.10 makes use of the logistic vertical scale that transforms S-curves into straight lines. It becomes evident from this picture that a century-long history of an energy type can be described quite well with only two constants, those required to define a straight line. (The curved sections are calculated by subtracting the straight lines from 100 percent.) The future trajectory of an energy source is decided as soon as the two constants describing the straight line can be determined.

There are other messages in Figure 4.10. By looking more closely at the data we see that world-shaking events such as wars, skyrocketing energy prices, and recessions have had little effect on the overall trends. Strikes may be more visible. In the coal industry, for example, such actions may result in short-lived deviations but the previous trend is quickly resumed.

Another observation is that there is no relationship between the utilization and the reserves of a primary energy source. It seems that the market moves away from a certain primary energy source long before the source becomes exhausted, at least at world level. This was true for wood and coal. It should also be true for oil. Despite the ominous predictions made in the 1950s that we would run out of oil in twenty years, we never did; more oil was found as the demand grew. Oil reserves will probably never be exhausted because other energy sources will be introduced in time. Well-established substitution processes with long time constants are fundamental in nature and will not be influenced by "lesser" reasons such as the depletion of reserves.

Environmentalists have opposed nuclear energy vehemently. This primary energy source reached one percent share in the world market in the mid 1970s. The rate of growth during the first decade, however, seems disproportionately rapid compared to the entry and exit slopes of wood, coal, oil and natural gas, all of which conform closely to a more gradual rate. At the same time, the opposition to nuclear energy also seemed out of proportion when compared to other environmental issues. Could it be that environmentalists did not react to nuclear energy per se but to its *rate of growth* instead?

As a consequence of the intense criticism, the growth of nuclear energy has slowed considerably, and has now approached the straight line proposed by the model. One may question what was the prime mover here—the environmental concerns that succeeded in slowing the rate of growth or the nuclear energy craze that forced environmentalists to react?

The coming to life of such a craze is understandable. Nuclear energy made a world-shaking appearance in the closing act of World War II by demonstrating the human ability to access superhuman powers. The word superhuman is appropriate because the bases of nuclear reactions are the mechanisms through which stars generate their energy. Humans for the first time possessed the sources of power that feed our sun, which was often considered as a god in the past. At the same time mankind acquired independence; nuclear is the only energy source that would remain indefinitely at our disposal if the sun went out.

Energy Consumption Worldwide

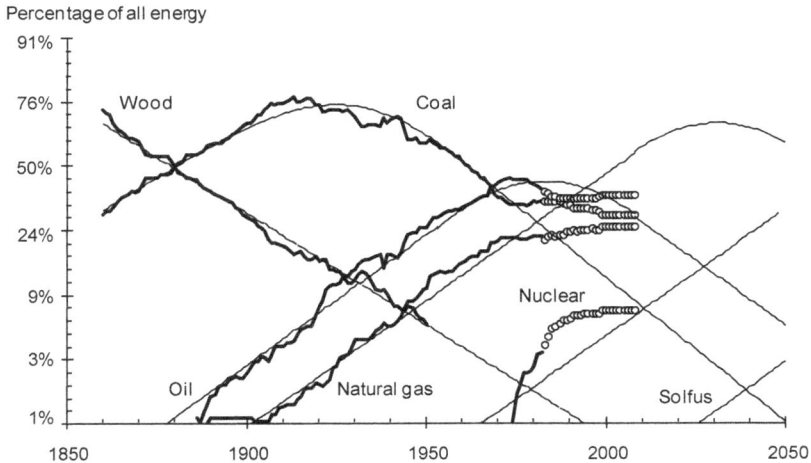

FIGURE 4.10 Data, fits, and projections for the shares of different energies consumed worldwide. For nuclear, the straight line is not a fit but a trajectory suggested by analogy. The little circles indicate recent data. The vertical scale is logistic. The futuristic source labeled "Solfus" may involve solar energy and thermonuclear fusion.[4]

The figure also suggests that nuclear energy has a long future. Its share should grow at a slower more *natural* rate, with a trajectory parallel to those of oil, coal, and natural gas. A more mature less hastened diffusion of nuclear energy will meet less resistance from environmentalists, if for no other reason the fact that a mature technology is less accident-prone. Indeed, in the last twenty years there has been less than one major accident in five years whereas in the early 1980s we witnessed five such accidents in three years.

The hypothetical primary energy source christened "Solfus" by Marchetti may include fusion and/or solar and/or other possibilities. It is projected to enter the picture in the 2020s supplying one percent of the world's needs. This projection is reasonable because whatever such a technology may consist of, once demonstrated to be feasible, would require about a generation to be industrialized, as did nuclear energy. But even if we had such an energy source available today, it would have to diffuse at the natural rate, the rate at which other types of energy have entered and exited in the past; otherwise it could meet opposition comparable to that from environmentalists to early nuclear. One way or another, the gas and nuclear cycles would still be traced out if somewhat earlier and smaller. Both these energy sources need to take their turn in playing a role comparable in importance to that of oil at its time.

But there is a significant "glitch" in the otherwise coherent energy picture of Figure 4.10. The share of coal stopped declining along the model's natural-growth trajectory in the early 1970s at the expense of natural gas. This may not be due to only the aggressively developing countries such as China who use coal ravenously. Developed countries such as the UK have also proven reluctant to give up coal. Whoever the culprit, the widening gap between the persistent level of coal use and coal's naturally declining trajectory becomes a source of pressure to the system, which may manifest itself in unexpected ways (possibly another case like the environmentalists vs. nuclear in the 1980s).

WHEN WILL HYDROGEN COME?

There is a secret concealed in Figure 4.10. As society moves from wood to coal to oil to gas and to nuclear society unwittingly pursues a strategy of fuel improvement, not only because each new fuel is cleaner fuel but also—not unrelated—because each new fuel has higher energy content. Wood is rich in carbon but natural gas is rich in hydrogen. When hydrogen burns it produces water as exhaust; when carbon burns it produces CO_2. When wood burns, very little hydrogen becomes oxidized to become water. Most of the energy comes from the carbon that oxidizes into CO_2. On the contrary, when natural gas burns, lots of hydrogen molecules become water and very little carbon becomes CO_2. The molar ratio hydrogen/carbon for wood is about 0.1, for coal about 1, for oil about 2, and for natural gas (e.g., methane) about 4. For a fuel like hydrogen this ration becomes infinite and the CO_2 emissions to the atmosphere null.[5]

Bio-fuels such as ethanol have a molar ratio of 3 and therefore on a quality basis they belong between oil and natural gas. Thus introducing bio-fuels on a big scale today would represent a move backward in the evolution of fuels in society.

The energy substitution described in Figure 4.10 took place in such a way that fuels rich in hydrogen progressively and consistently replaced fuels rich in carbon, and all that happened in a *natural* way (i.e., following an S-curve). The combination of energy sources, according to the shares shown in Figure 4.10, yields a hydrogen content that increases along an S-curve, see Figure 4.11. Society followed this S-curve on a global scale without the conscious intervention of governments or other influential decision makers. Bottom-up forces have safeguarded for one hundred years a smooth transition to energies that perform more efficiently and pollute less.

The black dots in the top graph of Figure 4.11 have been obtained using the data points in Figure 4.10. Coincidental with the "glitch" mentioned earlier, there is now a deviation from the S-curve pattern beginning around 1972. It seems that the hydrogen-enrichment process (decarbonization) stopped at that time. The persistent use of coal and its impact on natural gas, however, are not alone to blame for the missing hydrogen in our fuels today. Had coal continued declining and gas ascending along their natural paths, we would still be missing some hydrogen today.

The black dots in lower graph of Figure 4.11 have been obtained using the smooth trend lines, as defined by the substitution model in Figure 4.10. Here too, there is a deviation from the S-pattern beginning around year 2000. This is because there is no hydrogen content in nuclear energy or in solar/fusion. As a consequence, the deviation from the S-curve becomes progressively more pronounced toward year 2050.

The gray area in the figure represents the "missing" hydrogen content. This amount of hydrogen should somehow be contributed by nuclear energy, if we want to continue the well-established natural course of decarbonization. Nuclear energy can indeed do this in a number

Hydrogen Content of the Primary-Energy Mix (data)

Hydrogen Content of the Primary-Energy Mix (model)

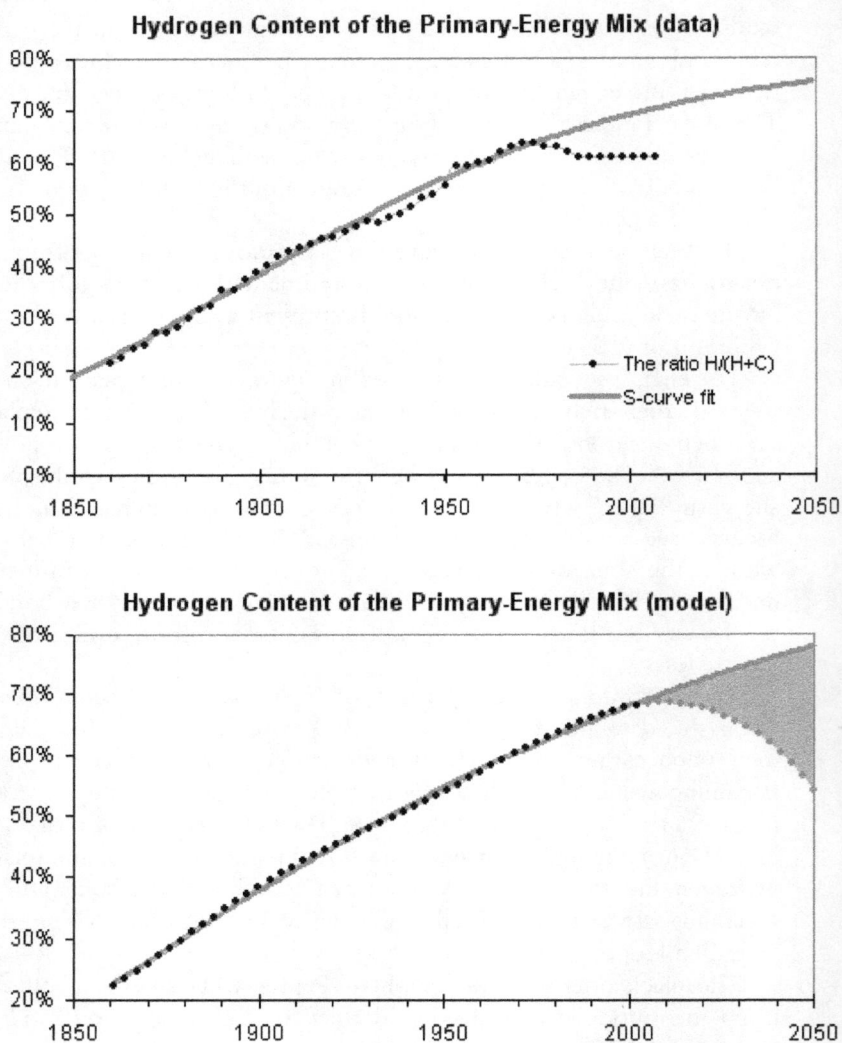

FIGURE 4.11 The black dots indicate the evolution of the hydrogen-content percentage according to the energy mix of Figure 4.10 from the data points (graph on the top) and from the model lines (graph on the bottom). The thick gray lines are S-curve fits to the black dots over the period 1860-1972 (graph on the top) and 1860-2000 (graph on the bottom). The gray area reflects the amount of hydrogen that needs to be provided by non-fossil types of energy.

of different ways. For example, seawater can be split into hydrogen and oxygen via electrolysis or by direct use of nuclear heat. It must be noted that nuclear energy is not indispensable for the natural path to be maintained. There can be other energy sources such as solar, wind, hydroelectric, thermonuclear fusion or a combination thereof, that could do the job, but these technologies are still responsible for insignificant contribution to the energy picture worldwide.

Moreover, some of these technologies—e.g., exploitation of wind energy and wave power—go against a yet another natural-growth process: the diminishing size of installations required to produce a certain amount of electricity. From Edison's time until today, power generators have increased in output but not in size, which emerges as another invariant. Be it coal burning, oil burning, or even a nuclear plant, the energy generator is generally housed inside a large building. In sharp contrast, equivalent amount of electricity produced by windmill farms or wave power would require thousands of square miles. Photovoltaic sun collectors would also require an area much greater than a large building. A successful manufacturer of photovoltaic rooftops in Germany has recently confessed to me: "The government leaves us no choice. The legislations from Brussels combined with the money the German government pours into green energies have created a demand, so we will sell into this market. But it is all money down the drain." A tennis-court size rooftop with voltaic cells integrated will produce 22 kilowatts of electricity; a drop in the ocean hardly justifying the endeavor.

These clean-energy alternatives may receive much visibility and public support but remain impractical, expensive and not very clean after all. An in-depth analysis reveals that the CO_2 emitted during the production and maintenance of these structures throughout their lifetime surpasses the CO_2 that would have been emitted by producing the same amount of electricity via burning oil. And there are more intricate sources of pollution; refining neodymium, a ton of which is in the average wind turbine, produces toxic and even radioactive wastes.

There is little doubt that society will eventually use hydrogen as its principle fuel because it is the most potent fuel and progress cannot be stopped. The use of hydrogen as fuel has demonstrated survival of "infant mortality" with extensive applications in rockets. It is only a question of time before it diffuses to other social uses. After all, no niche in nature was ever left partially filled under natural circumstances and an S-curve that has been evolving for one hundred years (Figure 4.11) will most certainly proceed to completion. The catch phrase here is "natural circumstances". Can we trust circumstances to remain natural? Figure 4.10 indicates an anomaly that qualifies as unnatural; coal consumption began deviating from its naturally declining trajectory in the early 1970s and is continuing to do so.

THE PRICE OF PRIMARY ENERGY

Another thing nuclear energy has going for it is price. If we consider only production costs (i.e., operations, maintenance, and fuel costs) the price of electricity produced from nuclear energy is the cheapest today.

The price of primary energy is somewhat of a sacred cow because energy is like food for society. Suddenly doubling the price of bread in a large country like India would produce a social uprising. The "steady-state" price of oil has more than doubled between 2006 and 2008 and yet there was no social uprising. Could it be that the price of oil has reached its mature level only recently?

A 200-hundred year chart—Figure 4.12—of the energy price shows an intriguing pattern. Huge spikes stand out about 56-years apart echoing the Kondratieff economic cycle (discussed next chapter). In between spikes, energy prices are confined to significantly lower levels.

These spikes are so pronounced compared with the usual day-to-day price fluctuations and are so regularly spaced that they inspire confidence in making some daring forecasts, for example that the next significant peak in the price of energy should take place in the late 2030s, not much earlier.

In the last fifty years, oil has been the major primary energy source and therefore the gray line in Figure 4.12 shows the price of oil. Because of the dollar's significant loss of value in recent years, oil prices are shown in Swiss francs, one of the world's most stable currencies.

The first three peaks in Figure 4.12, depicting Fuel & Lighting prices, are not all of the same size. What if the 1980 oil peak was more like the 1864 peak of Fuel & Lighting? That would imply that oil has generally been too cheap for the most of its existence. The background under the first spike at the left is almost at the height of the tip of the second spike. By analogy, oil's more recent high prices could be part of the background under a much higher spike, to be expected around 2036. Such a scenario has been sketched with the thin gray line to the right in Figure 4.12. In between spikes, a general steady-state price for oil should then be expected in the range 70 to 110 Swiss francs of December 2007.

The Price of Primary Energy

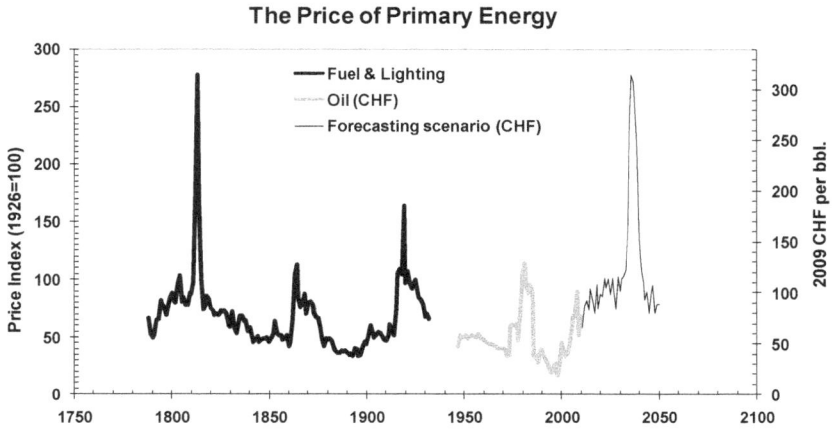

FIGURE 4.12 Average prices paid for energy in the US, corrected for inflation. The oil price, thick gray line depicted on the right vertical axis, is expressed in Swiss francs (CHF). The thin black line is a scenario for the future inspired by the pattern of the other four peaks.[6]

Nowadays this corresponds to roughly the same range in 2008 dollars, but the 2036 spike could correspond to well above $305 if the dollar declines further in the future.

WORLD ENERGY NEEDS

Per-capita energy consumption worldwide is more than seven times greater today than it was 150 years ago. This increase took place, not in a steady, uniform rate, or even in a random fashion, but in well-defined S-shaped steps.

Figure 4.13 was published in *Predictions* with data only up to 1985. At that time the third S-curve denoted by the intermittent line was only a hypothetical scenario for the future. But the data during the last twenty years (open circles) confirm this scenario as reality. Major economic crises such as the Lehman Brothers financial crisis and the European sovereign-debt crisis did not get in the way of the natural-growth evolution as anticipated by analogy with the previous two steps.

It must be pointed out that each one of the first two growth steps represents an increase of about a factor of two. A similar factor must be expected during the third step that just began. But a factor of only four since mid 19th century can hardly explain the abundance of work carried

Per Capita Annual Energy Consumption Worldwide

Tons of coal-
equivalent

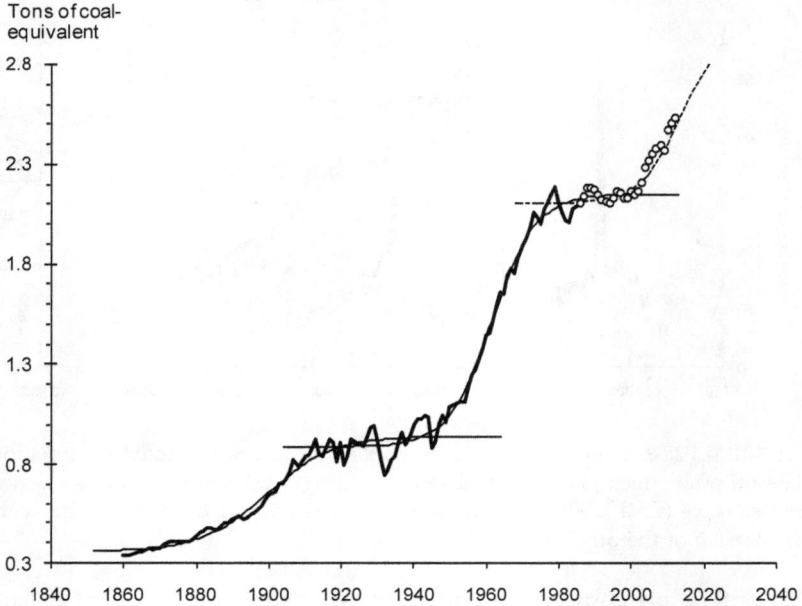

FIGURE 4.13 The data display sustained growth in terms of a succession of S-shaped steps. The two smooth solid lines are S-curve fits to the data. The intermittent line is a scenario for the future suggested by analogy. The open circles are recent data in remarkable agreement with the scenario proposed twenty years ago.[7]

out in society since then. What also increased during the same time are improvements in our efficiency while using the energy consumed.

Being 100 percent efficient is a God-like quality. In that respect humans still have a long way to go. In spite of the specific gains in efficiency over the last three centuries, if we take all energy uses together in the developed countries, we find that the overall efficiency for energy use today is about 5 percent. It means that to produce a certain amount of useful work we must consume twenty times this amount in energy. This result makes humans look shamefully wasteful.

Becoming more efficient is a learning process; consequently the evolution of efficiency is expected to follow the familiar S-shaped natural-growth curve. The evolution of efficiency for prime movers (engines), with its beginning around 1700 is shown in Figure 4.14. The S-curve fit (gray) is as determined in *Predictions*. The three recent data points, two of them predictions, are in good agreement with the S-curve.

The Evolution of the Efficiency of Motors

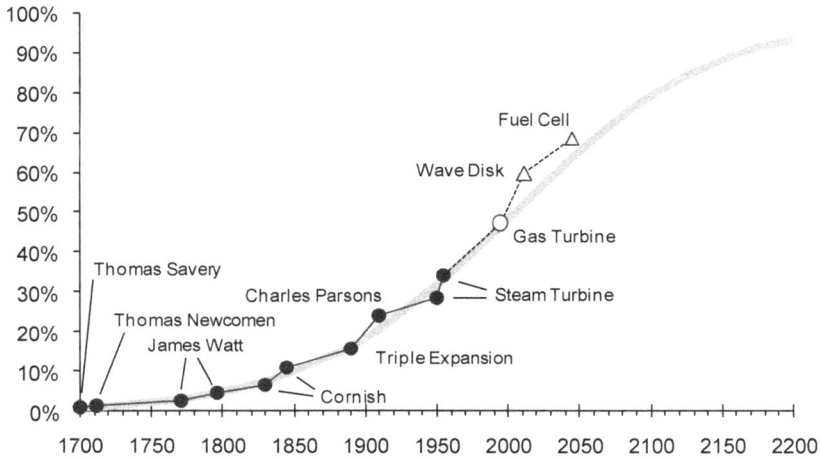

FIGURE 4.14 The S-curve fit (gray line) is on the data up to 1960 (black dots). The gas-turbine (open circle) represents more recent data. The two little triangles are estimates for future engines.[8]

IN CLOSING

At present the stress points to the social system are climate warming (carbon emissions), the price of oil, and food shortages. It is not obvious, however, what the origins of the problems are and what needs to be done. Stress is a symptom of interference with the evolution of a natural-growth process; the greater the interference the higher the stress.

One natural-growth process interfered with is decarbonization (moving toward energies with higher hydrogen content). Decarbonization has deviated from its natural-growth pattern in the 1970s and has stagnated ever since. The deviation (and consequently some of the stress) would diminish if deployment of nuclear energy were to begin increasing again at the natural rate and served to produce hydrogen. But that would not suffice. The excessive consumption of coal worldwide must also diminish in favor of more consumption of natural gas.

There is no shortage of oil, and high oil prices are not caused by production issues. In fact, oil at $100 per barrel (2008 dollars) may be a natural price. Price fixing, speculation, and warfare are not likely to raise the price of oil much above this level in the near future. The next manifold price hike should normally take place sometime in the mid 2030s.

The proposals for bio-fuels seem anachronistic. Bio-fuels not only waste food resources, they also yield less energy content and pollute more than natural gas. Our cars (and airplanes) should also be running on natural gas, as many municipal bus systems already do. Cars that use natural gas would be less polluting and much simpler and cheaper to build than hybrid cars. Increasing efficiency is good, but is not worth pursuing at any cost. Moreover, efficiency alone will never yield the factors of two and three that the world needs to grow in energy consumed per capita over the next fifty years, which is another natural-growth process in need to be respected.

5

Deviations from Natural Growth

"It turns out that an eerie type of chaos can lurk just behind a façade of order—and yet, deep inside the chaos lurks an even eerier type of order."

Douglas Hofstadter

• • •

FROM S-CURVES TO CHAOS

Multiplying rabbits may overshoot the niche capacity of the range they occupy while they explore the limits of how many of them can fit in it. This introduces a deviation from the S-curve pattern at the level of the ceiling. At the other end, the beginning of the S-curve, another deviation is the catching-up effect often encountered in the productivity curves of movie directors like Hitchcock and Fellini mentioned earlier (Figures 1.3 and Appendix Figure 4.5 in Appendix A respectively).

The early catching-up effect and the eventual overshoot of the final ceiling are both manifestations of the intimate relationship between S-curves and chaos. In scientific terms chaos is the name given to a set of ongoing variations that never reproduce identically. It was first observed when mathematical functions were put in discrete forms. Since populations are made up of discrete entities, a continuous mathematical function offers only an approximate model for the real situation. Discretization is also dictated by the need to use computers, which treat information in bits and pieces rather than as continuous variables. When the natural-growth equation that describes growth in competition is cast into discrete form it becomes the chaos equation. The latter is strikingly similar to the former—see Appendix D—but whereas the former gives rise to the smooth S-curve, the latter gives rise to states of chaos. The S-curve has become the tool to describe natural growth. A state of chaos denotes the lack of a growth trend and has become the tool to describe erratic fluctuations. Both pictures originate with growth in competition of a Darwinian nature (i.e. survival of the fittest). States of chaos appear on what corresponds to the ceiling of the S-curve after the upward trend has died down, see Figure 5.1.

It has also been shown that chaotic-type fluctuations could be expected before *as well as* after the curve's steep rise, a picture that would

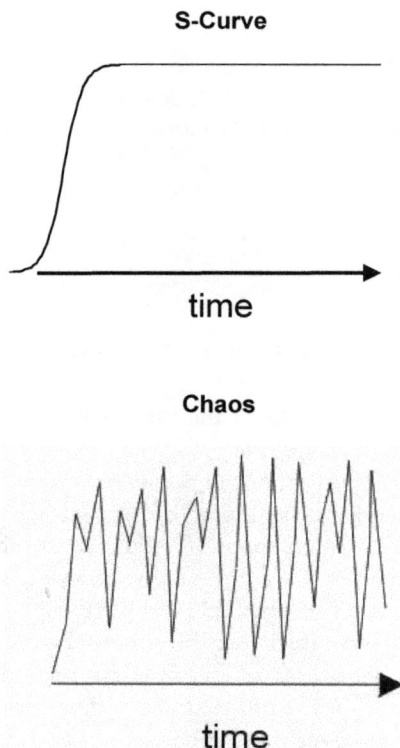

S-Curve

time

Chaos

time

FIGURE 5.1 At the top we see the solution of the natural-growth equation giving an S-curve. At the bottom is the corresponding solution of the chaos equation giving mostly erratic fluctuations.

arise from cascading S-curves when a new S-curve begins where the old one is completed.[1] A large time-frame such example is the world economy, as evidenced by the evolution of energy consumption.

Energy Consumption

As we saw in the previous chapter the worldwide energy consumption grew over the last two centuries along the pattern of three successive S-curve steps (Figure 4.13). It is not surprising that the low-growth periods, one around the 1920s and another one in the 1980s, correspond to economic recessions/depressions. The regular alternation between high-growth and low-growth phases again echoes Kondratieff's economic cycle to be discussed in next chapter.

It is easy to see why energy consumption is a competitive process. Human appetite for energy is insatiable—people will use up all the energy they can get—but supply is limited because the procurement of energy is difficult (read expensive). From time to time technology and socioeconomic conditions permit/stimulate the opening up of new energy-supply niches. When this happens energy consumption increases to exhaust these niches in a natural way, namely along S-shaped patterns. At the end of the growth step energy consumption reaches a homeostasis; further growth is held back by other more urgent priorities.

In Figure 4.13 we saw that the first step ended around 1920 with a period of stagnation that lasted for about two decades. The second energy consumption step was completed around 1975, and we have just witnessed the beginning of a third step. There can be little doubt that this indicator will go through another growth phase considering the dire need for industrial growth in the developing world.

We will come back to discuss the many ramifications of this figure in the next chapter. But here let's just pointed out that the evolution of this indicator unambiguously displays the alternation between growth and stagnation periods with the former characterized by "orderly" data points while the latter is characterized by chaotic fluctuations. At the ceiling of the S-curve a state of chaos becomes prominent and as theoretically expected such a state precedes as well as follows the steeply-rising growth phase.

The onset of these instabilities includes the precursor, the catching-up effect, and the ceiling overshoot, see Figure 5.2. In real-life situations we cannot see the early oscillation completely, because negative values have no physical meaning. We see, however, a precursor followed by a quiet period, then an accelerated growth rate, and finally an overshoot of the ceiling. These features correspond to real phenomena. Accelerated growth is a catching-up effect, usually attributed to pent-up demand. The overshoot is a typical introduction into the final steady state. As for the precursor, it is often considered a fiasco, unfairly so.

The Concord

We saw in Chapter 1 that the daily displacement invariant of seventy minutes combined with the speed of our transport defines the size of our cities. A supersonic airplane of the future of average speed, say 3,000 miles per hour, would render the whole Western world as one town. In his book *Megatrends*, John Naisbitt claimed that Marshall McLuhan's "world village" was realized when communication satellites linked the world together informationally.[2] This is not quite the case. Information exchange is a necessary but not a sufficient condition. It is true that empires in antiquity broke up when they grew so large that it took more than two weeks to transmit a message from one end to the other. But it is also true that communications media are poor substitutes for personal contact. The sufficient condition for a "world village" is that it should take not much more than one hour to physically reach any location in it.

Expanding in space as far as possible is of primordial importance for all species. Accordingly supersonic travel and the realization of "world village" become inevitable some time in the future. But there are constraints. Productivity (that is, the product speed times payload) cannot be compromised. As with the standard of living, decrease is not an option for a new way of life. The productivity of a new-technology aircraft must increase. This was not the case with the Concord whose productivity significantly fell short of the productivity of the Boeing 747 introduced around the same time. The Concord's payload was too small. It should be able to carry around 250 passengers to match the productivity of wide-body aircraft. And even with 100 passengers the time the Concord gained flying, it lost refueling. This explains its commercial failure but not the end of supersonic travel.

Aviation know-how has already achieved maturity today, and simple technological advances cannot produce factors of ten improvements. A new technology, fundamentally different, is required—for example, airplanes fueled with liquid natural gas or liquid hydrogen. The new competitor must possess an indisputable competitive advantage. The scrapping of the supersonic transport project (SST) back in 1971 by the

Senate displeased the Nixon Administration but may have been symptomatic of fundamental, if unconscious, reasoning. The comment by then Senator Henry M. Jackson, "This is a vote against science and technology," can be seen today as simplistic and insensitive to rising popular wisdom.

As we saw in the previous chapter supersonic aircraft technology will have to rely on a richer fuel, such as liquid natural gas or liquid hydrogen. This fuel will permit high productivity, namely, supersonic speeds as well as relatively high carrying capacity, but probably narrow fuselage (single corridor). Last but not least, the hydrogen-rich fuel will do marvels for the environment.

The Concorde enjoyed much publicity and popularity. It made a cultural dent and demonstrated the public's appreciation and need for high-speed travel. Its lifetime was limited for technical reasons. But it constitutes a precursor in the supersonic-travel growth process. The time scale in Figure 5.2 has been chosen accordingly. Following the Concorde there may be another 27 years or so with no commercial supersonic planes. However, once a new-fuel technology becomes available, growth in supersonic travel will be rapid because of pent-up demand. It may lead into a supersonic "craze" (overshoot) around 2060 *and finally stabilize at a lower level toward the end of the 21st century.*

Making an S-Curve Discrete

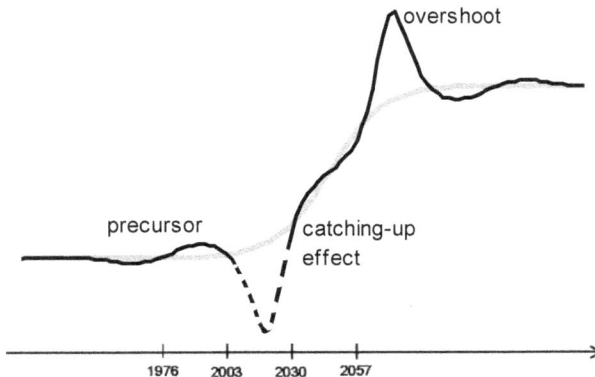

FIGURE 5.2 this is one of the first patterns we obtain by casting an S-curve in a discrete mathematical form. The deviations from the familiar S-curve demonstrate the precursor, the early catching-up effect, and the overshoot. The dates on the horizontal axis mark milestones for supersonic travel with the precursor designating the Concorde, see text.

The necessity for uniting the world combined with the fact that supersonic aviation technology has proceeded to beyond infant mortality—as demonstrated by the Concord, the Russian Tupolev Tu-144, and a host of military aircraft—guarantee that this growth process will continue to completion.

INSIGHTS INTO FLUCTUATIONS

When S-curves cascade they are separated by low-growth periods during which a state of chaos appears, reflecting turbulent times, see Figure 5.3. There is both theoretical and practical evidence for the appearance of chaotic fluctuations at the end and at the beginning of an S-curve.[3]

During one of my talks someone in the audience objected to my schematic representation of sustained growth as a sequence of cascading S-curves with interspersed chaotic intervals.

"There may also be chaotic fluctuations all along the rising part of the S-curves," he argued. "They are simply less visible because they are masked by the pronounced upward trend."

To drive his argument home the man walked up to the blackboard and drew an S-curve with a trembling hand imitating an old man whose hands are shaking. Indeed on the curve he drew there were fluctuations

Cascading S-Curves

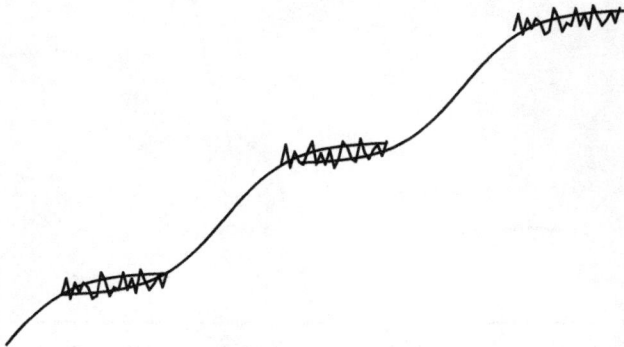

FIGURE 5.3   Sustained growth comes in well-defined steps and displays an alternation between states of order and states of chaos.

everywhere but they seemed more pronounced at the extremities of the S-curve.

While he was drawing on the blackboard I had a chance to prepare my answer. "Granted, there may fluctuations all along the S-shaped pattern," I admitted. "But there is a significant difference between fluctuations during the steeply rising part, and fluctuations at the top and at the bottom. During the fast-rising trend fluctuations can be considered 'benign' as they generally correspond to reality, whereas at the top and at the bottom they are truly chaotic as they will never become realized. Consider the fluctuations I will highlight with a little circle. I will indicate the corresponding level on the S-shaped pattern with a black dot," I said and proceeded to add little circles and black dots on the drawing on the blackboard, see Figure 5.4.

"During the steeply rising part, the distance between circle and dot is small, but it becomes bigger as we approach the ceiling. At the ceiling this distance becomes infinite. A fluctuation during the rising trend corresponds to a phenomenon that naturally appears either a short time earlier or a short time later. The only 'abnormality' introduced by such a fluctuation is that of a precocious or a belated appearance. But the same-size fluctuation at the ceiling reaches a level that will never be achieved by the natural-growth process, and thus has zero chance of being *naturally* realized. Such a fluctuation can be considered as unnatural.

Drawing an S-Curve with a Trembling Hand

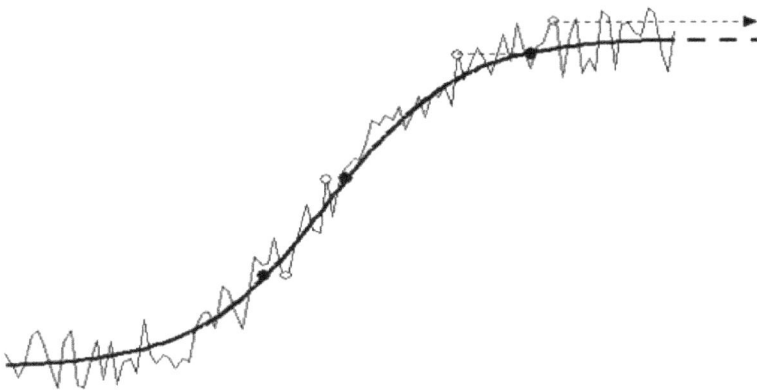

FIGURE 5.4 The fluctuations during the steeply-rising section are less visible because they are masked by the pronounced upward trend.

It would be simply *unusual* if it snowed in Greece in October, but it would seem *unnatural* if it snowed in the Sahara. During irregularities events occur earlier or later than their natural time, but during chaotic events they have no corresponding time in which they could seem natural."

Without realizing it, I had come up with a procedure for distinguishing between natural fluctuations (i.e., simply statistical fluctuations) or unnatural ones (i.e., chaotic). Naturalness is intuitively associated with the probability of being realized by the natural-growth process depicted by the smooth S-shaped pattern. A downward fluctuation at the ceiling would appear less and less natural the further it occurred from the S-curve. Even if the level of its extremity had been realized sometime in the past, if this was very long time ago, the fluctuation would seem rather unnatural. And this is the reason that chaotic fluctuations at the ceiling are to be considered as generally unnatural. They are mostly far removed from the initial rise to that level. Whether they are upward or downward they appear unnatural because a realization of this value via a natural-growth process is either impossible or it took place a very long time ago.

I was pleased with my answer. Once again S-curves had demonstrated solid common sense. Back home I looked again at the drawing of the S-curve by someone whose hands are shaking. I felt there was a way to quantify the common-sense conclusions I had drawn in my talk earlier that day. A natural fluctuation should somehow be connected to the bell-shaped life-cycle representation of natural growth.

In Figure 5.5 consider a fluctuation above the S-curve at time t_1 during the curve's steep rise. The extremity of this fluctuation corresponds to the curve's level at time t_2. This fluctuation is natural because the level reached will appear in a short time in the future. But the same fluctuation at time t_3 reaches a level that has zero chance of ever being realized. Such a fluctuation can be considered as unnatural. Superimposing the bell-shaped pattern of the natural-growth life cycle on the same drawing revealed the rule I was looking for.

Whenever the fluctuation reaches a level that corresponds to a time within the gray bell-shaped area, the fluctuation can be considered as natural. At time t_3 the fluctuation should have much smaller amplitude to qualify for naturalness unless it was in the downward direction. For a fluctuation to be natural its extremities must correspond to points on the curve that map to the gray bell-shaped pattern, which is the natural-growth life cycle.

Two Types of Fluctuations

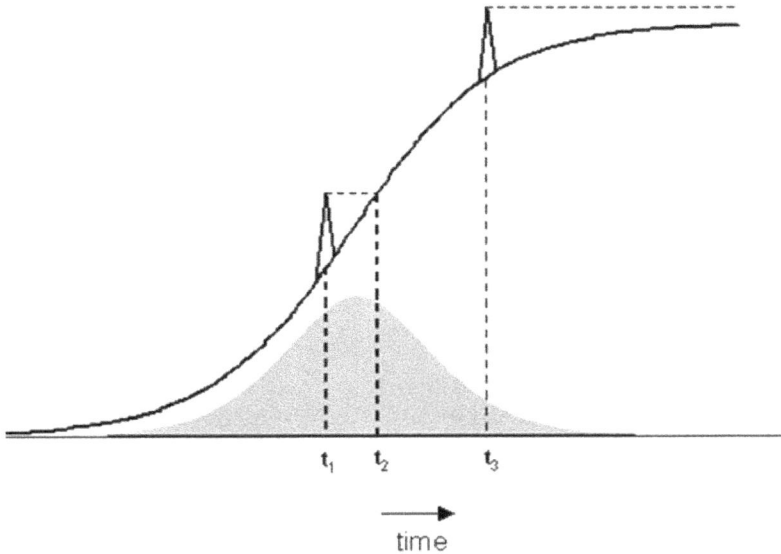

FIGURE 5.5 A fluctuation at t_1 reaches the same level as the logistic curve at t_2. The same-size fluctuation at time t_3 has no corresponding point on the curve. The gray life cycle delimits the position and size of all natural fluctuations. If the fluctuation at t_3 was much smaller, it could have been natural.[4]

S-CURVES WITH VARIABLE CEILING

An S-curve determined through a fit to a series of data points will have a tendency to flatten toward a ceiling as early and as low as it is possible within the constraints of the fitting procedure. Therefore curve-fitting software programs will often yield S-curves that are biased toward a low ceiling. Uncertainties on the data accentuate this bias by permitting larger margins for the determination of the S-curve parameters. In fact the larger the fluctuations on the data the greater this bias may be. But it can be that this bias is unjustifiably blamed for, as is the case of US Nobel laureates described below.

It was Marchetti who first suggested that the competition for Nobel-Prize awards can be described by S-curves.[5] Following his example I tried to study the evolution of the number of Nobel Prizes won by the United States.

Winning a Nobel Prize is a competitive process because Nobel Prizes are desirable and at the same time they constitute a "limited resource" with a restrained number of them being awarded each year. By definition, the best-fit candidates win. Obviously, a peace Nobel Prize is very different from a prize in Physics. Moreover, some prizes may be shared among as many as three individuals whereas others are given to only one individual. Nevertheless for my study I counted each laureate as one independently of what discipline he or she was in and independently of how many colleagues shared the prize. The justification for this is that we are counting individuals with exceptional contributions to the benefit of mankind and on the average relative underachievements are compensated for by relative overachievements.

My first attempt to fit an S-curve to the cumulative number of US Nobel laureates in 1988 concluded that the US Nobel niche was already more than half full and implied a diminishing annual number of prizes for Americans from then onward.[6] Ten years later I confronted those forecasts with more recent data in my book *Predictions – 10 Years Later.*[7] The agreement was not very good. The forecasts fell below the actual data and despite the fact that there was agreement within the uncertainties expected for a 90% confidence level the discrepancy did not go unnoticed. A technical note published in *Technological Forecasting & Social Change* in 2004 highlighted the inaccuracy of my forecasts and cast doubt in the use of S-curves to forecast US Nobel laureates.[8] On my part, I refit the updated data sample with a new S-curve pointing to a higher ceiling. I also began wondering whether there was evidence here for the known bias of S-curves to underestimate the final niche size despite my compensatory procedures. The new forecast again indicated an imminent decline in the annual number of American Nobel laureates.

Another ten years later I once again confronted my older forecasts with recent data. The situation turned out to be the same as ten years earlier, namely the forecasts again underestimated reality and despite agreement with the result of ten years earlier—within the uncertainties expected for a 90 percent confidence level—there was now clear disagreement between recent actual numbers and the original forecasts of twenty years earlier. The situation was reminiscent of the celebrated Michele-parameter episode in experimental physics where a measurement repeated many times over the period of fifty years kept reporting an ever-increasing value always compatible with the previous measurement but finally ending up in violent disagreement with the very first measurement.

One explanation for the S-curve ceiling to be constantly increasing is the fact that the US population itself has also been increasing over the same historical period. An increasing population provides an increasing "niche" for Nobel-Prize winners. But the mathematical equation that describes natural growth (see appendix B) can be analytically solved only if the niche capacity remains constant throughout the growth process. An S-shaped pattern for the US Nobel-laureate population is presumptuous and probably wrong if the niche capacity increases with time.

An obvious way to account for the growing American population would be to study the number of laureates *per capita* thus rendering the constant that describes the ceiling of the S-curve time-independent and the growth equation solvable. So I repeated the previous analysis this time for the number of Nobel Laureates per 100 million of inhabitants. I obtained different results with better-quality fits and consistency, namely the values of the ceiling for all three periods were within the expected uncertainties of ± 20% from each other. But there was still some tendency for the ceiling to increase—if within the estimated uncertainties of the results—and I also obtained counterintuitive forecasts for a dramatic decline of American Nobel laureates and/or a major increase of the American population by the second half of the 21st century.[9]

The tendency of S-curve ceiling to still grow with time suggested that considering US Nobel laureates *per capita* did not fully account for the increase of the "niche" size over time. In fact, the niche of individuals qualified for Nobel-Prize candidature in America could be increasing faster than the average population. After all, in my study I classified laureates with double nationality as nationals of the nation where the research for which they were being distinguished was accomplished. America, as a rule, welcomes research scientists from all over the world while it thwarts immigration by the uneducated. It could very well be that the population sample capable of producing Nobel laureates in America is growing faster than the rest of the population.

The need then arises for solving the natural-growth equation with the niche capacity a function of time, which can be done only numerically. The procedure is described in Appendix B. The results are plotted in Figure 5.6. With data up to 1988, we see my original S-curve fit as well as a fit with an S-curve of variable ceiling. The actual numbers during the following twenty years—depicted with little circles—confirm the suspicion that indeed we are dealing here with a niche that is growing faster than the population. The results also include annual forecasts that remain rather flat for the next few decades.

US Nobel Laureates per 100 Million Inhabitants

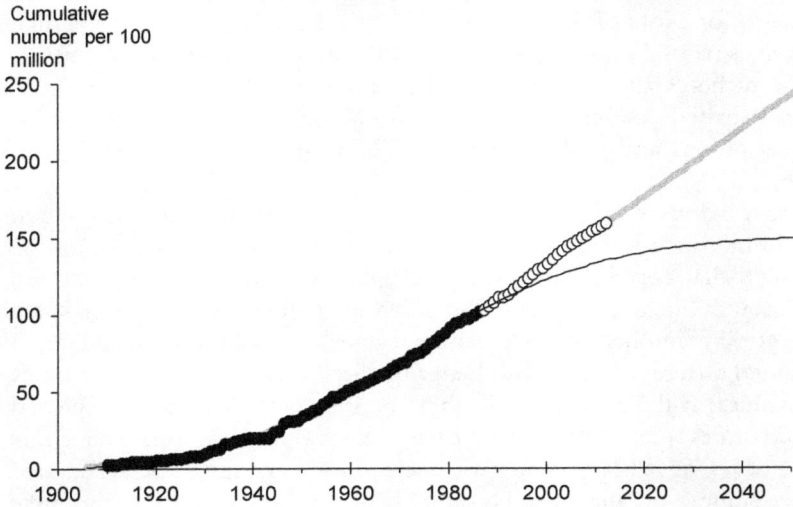

Cumulative
number per 100
million

Life Cycles

Annual number
per 100 million

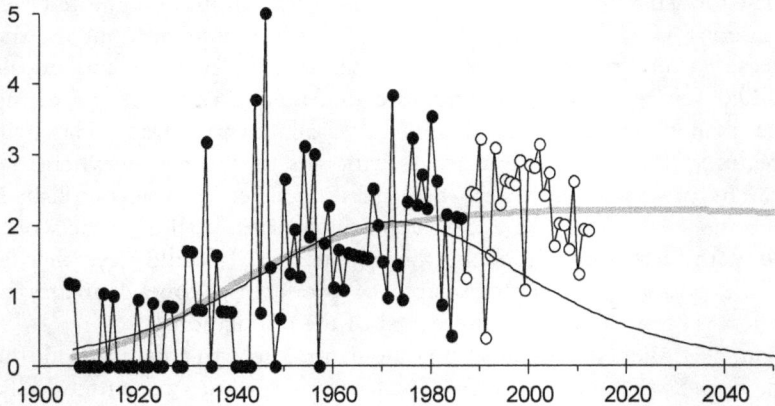

FIGURE 5.6 At the top we see the cumulative number of US Nobel Laureates per capita. The thin black line is an S-curve fit and the thick gray line a fit to an S-curve with variable—increasing with time—ceiling. The fits were performed on data up to 1988 (black dots). The little circles denote more recent data. At the bottom we see the life cycles that correspond to the curves at the top.

The S-curve with a variable ceiling is a rare deviation from the classical S-curve formulation. For most practical cases it is safe to assume that the niche capacity will remain constant over time.

S-curves are intimately related to chaos and generally speaking S-curves are important in modeling the stage of growth, whereas descriptions of chaos are more appropriate for the irregular fluctuations observed in the absence of growth. There are phenomena for which the initial rise in the growth pattern becomes quickly irrelevant—for example, building up a smoking habit. Every smoker begins from zero, but people are concerned with how much they smoke per day rather than the detailed path along which their habit was formed. The number of cigarettes smoked per day displays chaotic fluctuations. On the other hand, when it comes to the growth of a child's height, all the interest is focused on the S-curve traced by the measurements of the child's height over time; any small fluctuations after adolescence, when the main growth process is completed, are of no consequence whatsoever.

A well-established S-curve will point to the time when chaotic oscillations can be expected, namely when the ceiling is being approached. In contrast, an entrenched chaos will reveal nothing about when the next growth phase may start. In fact, a state of chaos can be considered as a bifurcation period following which an upward growth phase is as realistic as a downward one.

6

The Kondratieff Cycle

"If winter is here, can spring be far behind?"
Percy B. Shelley

• • •

Energy consumption correlates in an unambiguous way with industrial development and economic prosperity. The energy-consumption curve we saw earlier (Figure 4.13) points out two low-growth periods a couple of decades long each, one centered on the mid-1930s and another one around 1990. They are characterized by fluctuations resembling states of chaos. These economic winters are flanked by periods of growth of comparable length, one centered on the turn of the 20th century and the other around the 1950s and 1960s. These periods of economic boom are characterized by a rather orderly evolution of the growth pattern.

A commonsense explanation of the order-chaos alternation associates economic summer seasons with conservatism and tight control, in the spirit of not changing something that works well. At the same time, during economic winters, erratic trial-and-error searches abound, as means for finding new growth opportunities, and give rise to chaotic fluctuations. A four-seasons metaphor has been developed at length in my book *Conquering Uncertainty* because business, like the weather, goes through seasons, and so do the correct management policies.[1] The mechanisms associated with nature's four seasons can guide management decisions on business seasons. For example, the low creativity observed during summer is only partially due to the heat. New undertakings are disfavored mainly because summer living is easy and

91

there is no reason to look for change. In contrast, animals (for example, foxes and sparrows) are known to become entrepreneurial in the winter. There is wisdom encoded in nature's seasonal patterns and behaviors. As we will see later in this chapter these behaviors can be studied and transferred to whatever situation depicts a succession of seasonlike stages.

On the scale of Figure 4.13 the cadence of economic booms and busts resonates with the Kondratieff cycle. In 1926 the Russian economist Nikolai D. Kondratieff established evidence for a long economic wave with a period of 50 to 60 years.[2] He deduced this cycle from economic indicators alone. His work was promptly challenged. Critics doubted both the existence of Kondratieff's cycle and the causal explanation suggested by Harvard scholar Joseph A. Schumpeter. The latter tried to explain the existence of such cycles by attributing growth to the fact that major technological innovations come in clusters.[3]

Kondratieff's postulation ended up being largely ignored by contemporary economists for a variety of reasons. In the final analysis, however, the most significant reason for this rejection may have been the boldness of the conclusions drawn from such ambiguous and imprecise data as monetary and financial indicators, labels unreliable for assigning lasting value. But I have strived to provide evidence for a long cycle based on *physical* indicator such as per capita energy consumption expressed in tons of coal-equivalent. Such an approach was first adopted by Hugh B. Stewart.[4] If we visualize a large-scale S-curve superimposed on the data of Figure 4.13 to describe the overall trend, we will find the data wandering above and below this trend in a rather regular cyclical way. Figure 6.1 below extracts this cyclical variation by graphing the ratio of data to an overall S-curve trend. A rather regular variation emerges with a period of 56 years. The figure also shows the three most important stock-market crashes because they seem to be synchronized with this variation.

In 1873 the stock market in Vienna crashed at the heart of Europe whose economy dominated the world (the Dow Jones had not yet been conceived) and 1873 is exactly fifty-six years before 1929 when New York's stock market crashed. It was natural to ask at this point whether another crash took place fifty-six years further upstream, namely around 1817. It turns out that Napoleon's defeat at Waterloo in June 1815 touched off a speculative boom that ended in credit collapse and failures by autumn. At that time the state of the economy was not best represented by activities at the rudimentary stock exchange of that time. But commodity prices declined across the country, money tightened, and many country banks failed. By 1816 England was in a deep recession that lasted through 1825.[5] The post-Napoleon crisis corroborates the observation that Kondratieff's cycle may clock economic crises.

Energy Consumption Departed from an S-curve in a Cyclical Way

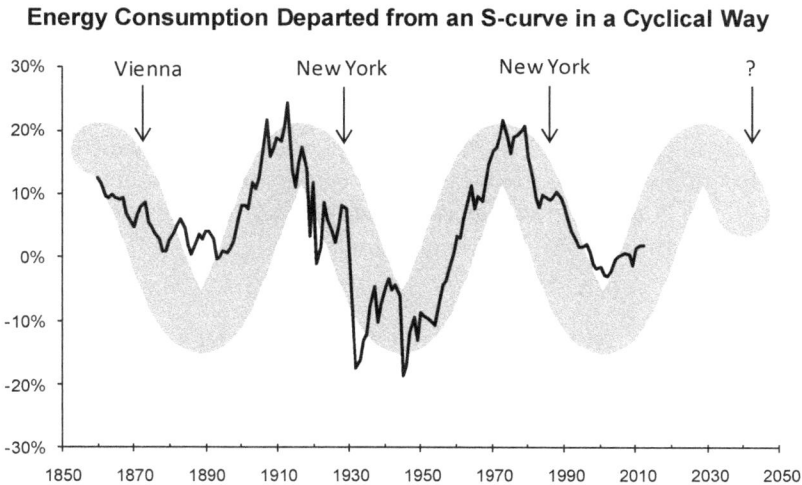

FIGURE 6.1 The data points represent the percentage deviation of per capita energy consumption worldwide from the natural growth-trend indicated by a fitted S-curve. The thick gray band has width of 8 percentage points around a regular variation with a period of 56 years. The arrows point at major stock-market crashes. The post-Napoleon economic crisis of 1815-1816 is not shown in the graph.

A similar graph in *Predictions* evidenced the Kondratieff cycle from energy consumption only in the US. That image is in full agreement with the one above because the data up to 1988 had finished on a downward trend toward the minimum and turnaround point in year 2000. Worldwide data are chosen here in order to directly relate to Figure 4.13 and to correlate with stock-market crashes around the world.

In *Predictions* I had also documented many other cyclical phenomena resonating with this energy-consumption cycle. For each human endeavor considered (production, consumption, manufacturing, construction, creativity, productivity, criminality, and the like) I used data expressed in their proper units, not by their prices. Through all these observations I was able to determine this cycle's period to be equal to fifty-six years, give or take two years. In Appendix A the reader will find updates of these *Predictions* graphs with recent data superimposed.

Several theoretical explanations have been offered for the origin of a long economic wave. In the 1930s Joseph A. Schumpeter at Harvard put forth socioeconomic theories, such as the rapid growth of leading sectors and the clustering of technological innovations. More recently, Jay Forrester at MIT was able to reproduce the same long wave with his sophisticated system-dynamics model, which studied major shifts in private-sector incentives for investing in capital plant, borrowing, and savings. Finally, in *Predictions*, I have offered two more hypotheses for the existence of such a long wave. One has been suggested by Cesare Marchetti and has to do with periodic changes of the climate. The other one has been suggested by Nobel-Prize laureate Simon van der Meer and has to do with the mean life span of a person's commercially active career. The connection to climate is been further supported here by new studies presented in the following two sections.

ATLANTIC TROPICAL HURRICANES

Tropical storms over the Atlantic seem to be more important and more frequent every year. During 2010 (a more or less typical year for hurricanes) out of 19 tropical storms 5 had winds more than 110 miles per hour thus qualifying for hurricane Category 3 or higher. In contrast, during the first five years of the 20th century in a total of 37 tropical storms (an average of 5 per year) there were only 2 hurricanes of Category 3. Coupled with recent facts on global warming and climate change these observations make one wonder what to expect in the years to come.

So I decided to look at this evolution of hurricanes from the point of view of a natural-growth process. In other words, as if some "natural" cause provokes this increase in frequency and strength of tropical storms and consequently the phenomenon should go through the stages of growth, maturity, and decline. I obtained detailed data on hurricanes since 1851 from the Tropical Prediction Center Best Track Reanalysis.[6] I retained data only about hurricanes of Category 3 and higher and grouped them together in buckets of decades. I was not surprise to discover not one but three natural-growth steps (smaller S-curves cascading). Figure 6.2 shows the three corresponding life cycles with peaks on years 1899, 1955, and 2011 respectively, while the maximum number of hurricanes per decade practically doubled between 1900 and 2010.

It is not obvious what phenomena are responsible for the progressive increase of the number and intensity of hurricanes over the last 150 years but global warming could well have something to do with it. But it is even less obvious why there are three cycles of about 56 years long.

Two things are clear. First that we are going to see more hurricanes per year in the years to come until the mid 2010s after which date their number may begin to decrease for a while; second that this ebb and flow of hurricane waves resonates with the Kondratieff cycle. This observation

Hurricanes: Ratio to an S-curve Trend

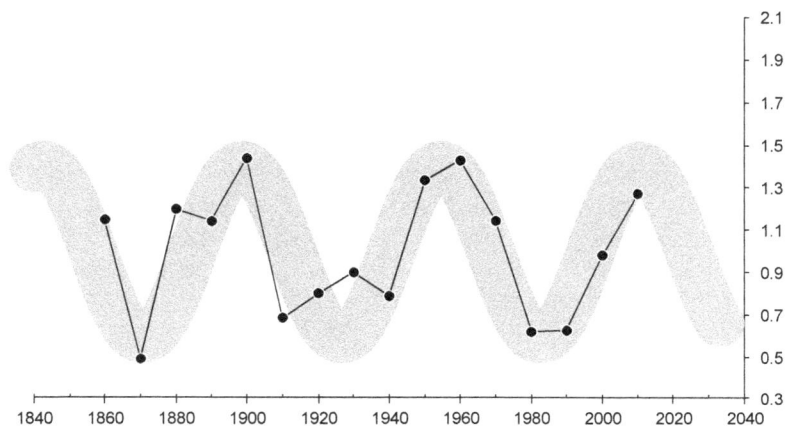

FIGURE 6.2 The data points show the percentage deviation of the number of hurricanes per decade from an overall S-curve trend. The thick gray band is a sinusoidal wave with period 56 years (Kondratieff's cycle) and width 5 percentage points.

gives support to the hypothesis that the origin of Kondratieff's cycle is climatic change. The thick gray band in Figure 6.2 is the Kondratieff's cycle as determined in Figure 6.1. There is excellent agreement between the hurricane data and this cycle as determined by the energy-consumption variation earlier.

SUNSPOTS

For centuries astronomers have been studying spots on the surface of the sun. Cooler than the rest of the sun's surface, these spots can last from a few days to many months, releasing into space huge bursts of energy and streams of charged particles. The effects of sunspot activity are varied and continue to be the object of scientific study. What is known is that they are of electromagnetic nature and that they perturb the earth's magnetic field, the ionosphere, and the amount of cosmic radiation falling on earth from outer space.

We also know that there is a regular eleven-year variation in sun-spot intensity. Sunspot activity reached a maximum again in 2013, as it had done in 2002, 1991, and 1980. During maximum activity, the overall solar output increases by a few tenths of 1 percent. The corresponding temperature change on the earth may be too small to be felt, but meteorologists in the National Climate Analysis Center have incorporated the solar cycle into their computer algorithms for the monthly and ninety-day seasonal forecasts. Every fifth period (five times eleven equals fifty-five years) the timing of the sunspot variation will be close enough to resonate with the Kondratieff cycle. Moreover, in the three-hundred-year-long history of documented sunspot activity, I have succeeded in detecting relative broad peaks in the number of sunspots every fifth period. Only one peak is missing around 1900. This observation was first published in *Predictions* and is confirmed below (Figure 6.3) with recent data.

EARTH'S TEMPERATURE

It is common knowledge that Earth's temperature has been rising in recent decades. What is less known is how the rate of this increase may be varying over time. Long-term temperature data come from Twin Cities, the Minneapolis–Saint Paul metropolitan area in east central Minnesota. Minneapolis and St. Paul, together known as the Twin Cities, is the core of the 15th largest metropolitan area in the United States. With a population of 3.6 million people, the region contains approximately 60% of the population of Minnesota. Due to its location in the northern and central portion of the US, the Twin Cities has the

A Slow Cyclical Variation in Sunspot Activity

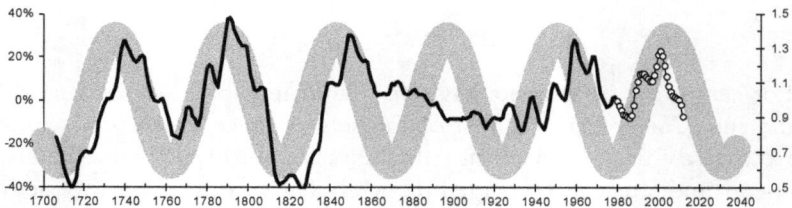

FIGURE 6.3   The graph shows the variation in the number of sunspots smoothed over a rolling 20-year period with respect to a 56-year moving average. Such a procedure—routinely used in time series analyses—washes out the well-known 11-year cycle of sunspot activity and reveals a longer periodic variation similar to the energy-consumption cycle. With one exception—a peak missing around 1900—the oscillation conforms best to a 54-year cycle shown by the thick gray band. The little circles indicate data from the last 20 years.[7]

96

coldest average temperature of any major metropolitan area in the United States.

Twin Cities is one of few locations across the central or western United States that have a continuous record of weather observations reaching back as far as 1820. Minneapolis and St. Paul have had early settlers who had the foresight to record the weather they experienced day to day for future generations who would call this area home. They created the database of weather records that helped to establish the climate of the region and began a tradition of meteorological science in the Northern Mississippi Valley.

Figure 6.4 shows the variation of temperature in Twin Cities with respect to a long-term S-curve trend. Despite the fact that the amplitude of this variation is small, it is rather regular and it unmistakably tracks the Kondratieff 56-year cycle.

The hypothesis that Kondratieff's cycle has its roots at the climate has been reinforced in this chapter by the observations that hurricanes, sunspot activity, and temperature variations have been resonating in tune with this cycle over hundreds of years. The reader must also take into consideration the other arguments made in *Predictions*, namely the cyclical appearances of lunar and solar eclipses with the same frequency, the regularly spaced steps on the continental shelf, and the darker circles that have been observed on centenary tree cross-sections with comparable periodicities.[9] All this points to environmental changes modulated by a 56-year pulsation.

Temperature Variation with Respect to a Growing Trend

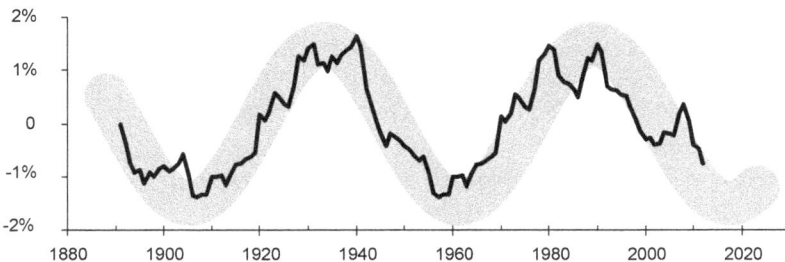

FIGURE 6.4 The black line shows the variation in temperature with respect to an S-curve trend, smoothed with a 22-year moving average. Such a procedure—routinely used in time series analyses—washes out the fluctuations and reveals a longer periodic variation tracking the energy-consumption cycle (thick gray band with a period of 56 years).[8]

It is not then unreasonable to suppose that human activities will follow suit. In fact, in their book *Climate and the Affairs of Men* Nels Winkless III and Iben Browning go into detail on how climate change has affected history.[10]

SEASONS OF GROWTH

The Kondratieff's cycle just like any other regular wave can be seen as a succession of life-cycle growth patterns. These patterns render themselves to segmentation into growth phases according to the rate of growth. Many management theorists divide growth cycles—typically products' sales cycles—into segments. They generally consider four segments according to the phase of growth: start-up, rapid growth, maturation, and decline. Their treatment is invariably qualitative, and the four phases are not necessarily of equal duration or precise definition. Here I will analyze the growth cycle somewhat differently, using the four seasons as a metaphor in order to reveal insights into behavioral patterns and deduce strategy-setting guidelines.[11]

Products, companies, and entire industries experience weather-like variations as do agricultural crops; they go through seasons in a cyclical way. Summer is the high-growth period around the midpoint of the cycle. Winters are the low-growth periods one finds at the end and at the beginning of the cyclical process. Between winter and summer comes spring, characterized by a progressively rising growth rate. Fall is the time between summer and winter, when the rate of growth continuously declines.

Borrowing images from biology to fit the marketplace is not new. Companies resemble living organisms. They are born; mature; get married; have daughters; become aggressive, sleepy, or exhausted; grow old; and eventually die or become prey to a voracious predator. Such waves are echoed in the preachings of management consulting gurus, who may pass easily from thesis to antithesis, and do not stop short of giving contradicting messages. It is not rare to see a company investing in business process reengineering and total quality—zero defects—at the same time. It is not trivial how to reconcile the benefits of leadership with those of empowerment and self-managed teams. We have seen advocates of centralized control and vertical integration to be replaced by voices supporting business units, core competencies, and horizontal corporations and vice versa. Such changes do not reflect conceptual breakthroughs in the theory of doing business. They are simply reactions to the economic climate and its seasonal variation.

Throughout history, periods of bureaucracy and control interspersed by waves of innovation and entrepreneurship. Notorious bureaucracies, such as the Roman Empire and the British civil service, were preceded and followed by entrepreneurial eras, such as crusades and revolutions (both social and industrial). The way to do business has followed suit. Many have addressed the question of how organizational behavior evolves over time. But it was Niccolo Machiavelli—early in the 16th century—who first pointed out the importance of adaptation. He wrote in *The Prince*:

> I believe also that he will be successful who directs his actions according to the spirit of the times, and that he whose actions do not accord with the times will not be successful.

A winter business season reflects the critical growth period encountered during the beginning and the end of a natural growth process. Products experience two winters in their lifetime. The first winter is while they are struggling for a foothold in the marketplace, and the second one when they are exiting and the follow-up product is fighting for succession. By definition, the end of the first winter signals that the growth process has survived "infant mortality"—that is, it has realized around 7% of its growth potential.

The seasons metaphor has more than poetic justification. The advantage over more traditional segmentations is that our familiarity with the mechanisms associated with nature's four seasons can shed light on and guide us through decisions on business and social issues. For example, the low creativity observed during summer is only partially due to the heat. New undertakings are disfavored mainly because summer living is easy and there is no reason to look for change. In contrast, animals (for example, foxes and sparrows) are known to become entrepreneurial in the winter. There is wisdom encoded in nature's seasonal patterns and behaviors. These can be studied and transferred to whatever situation depicts a succession of season-like stages.

Like the four seasons, the segments into which we divide the cycle must be of equal length. The time scale may vary widely depending on what growth process we are looking at. For a product, a season may last six months or a year. For an industry, a season may be five to ten tears. For the world economy cycling through 50- to 60-year waves, a season may be fifteen years long.

A product's first winter coincides with the incumbent product's second winter. Consequently this timing also implies that the new product must be launched during the fall season of the product it is replacing (no wonder farmers sow in autumn). Winter then becomes the time of selection, when wanton death eliminates the weak and the unfit. Spring corresponds to "adolescence", the formative years. Spring is also the time for research and development of future replacements.

Most product managers have intuitive knowledge of this product-succession sequence. They know, for example, that the new product is promoted most heavily while the old product phases out—to be precise, during the last 20% of its life cycle, that is, its second winter. They also know that research and development for the new product must parallel capacity buildup for the old product, which is approaching maximum rate of sales and profitability.

Figure 6.5 shows one business cycle, namely one S-curve growth step straddled between the end of the previous one and the beginning of the next one. The season assignment is such that summer occurs between 30% and 70% penetration of the S-curve and straddles the maximum rate of growth.

"Seasons" can be defined not only for products but for anything that grows in competition: markets, technologies, industries, and so on. It is generally true that spring is concerned with the *what* and fall with the *how*. That is why at the industry level product innovation occurs in spring and process innovation in fall. At the same time, at the economy level, technology and finances dominate in spring, and social and political forces dominate in fall. Spring is the time for investments. It is also the time for learning and continuous improvement. Specialists are in demand. Not so in winter. Sometime in the early 1990s I explained to a Geneva bank director that winter is the time to fire bureaucrats and hire Leonardo da Vincis—that is, cross-disciplinary, well-rounded men and women who stand a better chance than specialists to come up with revolutionary ideas for new profitable business. "Fire bureaucrats is exactly what we need to do, sir", he exclaimed. "Could you please tell us how to do it?" To my surprise, I heard two months later that the man had been fired.

But often what naturally happens is what should happen. As strange as it may sound, seeing your specialists progressively evolve into bureaucrats may be a good sign. It is one indication that summer is setting in. The word bureaucrats carries a negative connotation, but if we call them process agents instead, we realize that they provide an important function during times of high growth and prosperity. It is during summer that enterprises become successful, centralized, conservative (no one tampers with something that works well), and in need of clockwork operations. Fine-tuning and zero defects (the original

Assigning Seasons to the Life Cycle

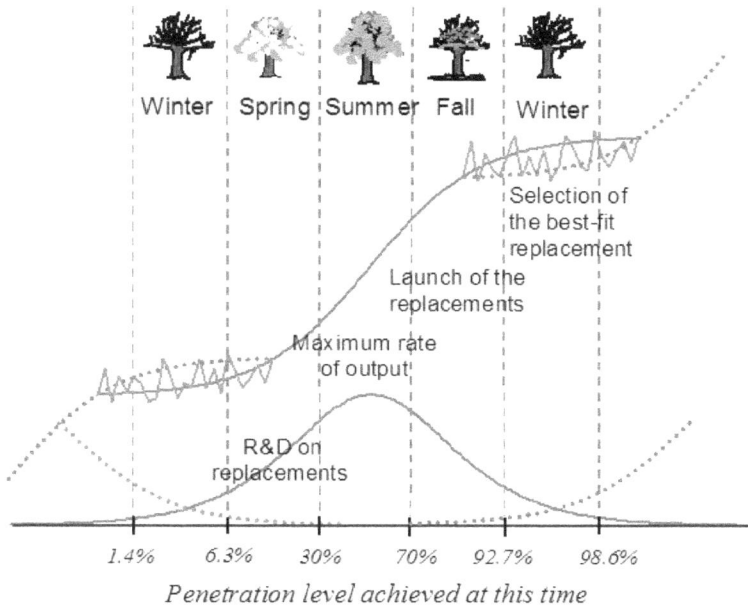

FIGURE 6.5 Segmentation of a business growth cycle into four seasons. The growth levels at the delimiting positions satisfy the following two conditions: (1) all seasons have the same duration, and (2) the early winter of the replacement overlaps with the late winter of the incumbent. Low-growth periods are accompanied by large and chaotic fluctuations.

aspiration of total quality management) are particularly appropriate for a summer season. But then, what about benchmarking, continuous improvement, and BPR (business process reengineering)?

Being second best hardly yields a competitive advantage. But positive feedback theories, that produce rapid fluctuations resembling chaos, argue that early gains for two simultaneously launched competitors eventually tilt the balance in favor of the "lucky" one and not necessarily the better one.[12] Early gains do not presuppose excellence.

When videocassette recorders were first introduced, the market was split between VHS and Beta. The two market shares fluctuated early on because of circumstances, luck, or marketing tactics. But soon early returns tilted the unstable situation toward VHS despite claims that Beta was technically superior. There are many such examples. Connoisseurs of personal computers value Apple products more highly than PCs, but the market-share gains of the latter have biased standardization in their favor.

Such manifestations of positive-feedback mechanisms have long been understood. During the nineteenth century Alfred Marshall—professor of political economy in Bristol, England—wrote that whatever firm first gets a good start will corner the market. To get a good early start, a product must appeal to the masses rather than to the elite, and that argues for postponing sophistication and refinements for a later season. New products are launched in the fall, but excellence is only excellent in the summer.

It is worth looking in more detail at each season's characteristics and how they can help us on everyday work decisions. There are advantages and disadvantages to each season. In my book *Conquering Uncertainty* I have devoted a chapter describing the characteristics of each season. One must keep in mind that these characteristics are meant to be in relative terms—that is, whatever happens in one season is with respect to what happened during the previous seasons. For example, to say that competition becomes lowest in spring does not mean it is negligible. It simply means that competition is relatively lower in spring than during the other seasons. Figure 6.6 lists some highlights for each season. It also

Highlights of Seasonal Recommendations

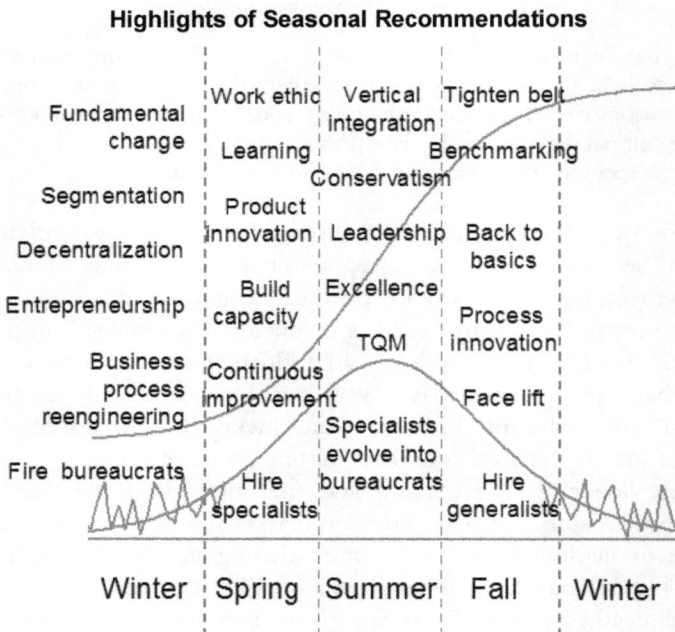

FIGURE 6.6 Highlights for strategies appropriate in each season.

sketches the chaotic fluctuations that accompany winters, which is an intrinsic characteristic of natural growth.[13]

S-curves enhanced by the seasons metaphor can also be used simply qualitatively to obtain rare insights and intuitive understanding. The seasons metaphor dictates that in spring the focus is on *what* to do, whereas in fall the emphasis shifts to the *how*. The former appears early in the growth process, the latter late. The evolution of classical music can be visualized as a large-timeframe S-curve beginning its development sometime in the fifteenth century and reaching a ceiling in the twentieth century. In Bach's time composers were concerned with what to say. The value of their music is on its architecture and as a consequence it can sound good when reproduced by any instrument, even by simple whistling. But two hundred years later composers such as Debussy wrote music that depends crucially on the interpretation, the how. Classical music was still "young" in Bach's time but was getting "old" by Debussy's time (when you hear people say that they need to focus on the how, you can understand that they are referring to something that is getting old,) see Figure 6.7.

The S-Curve of Classical Music

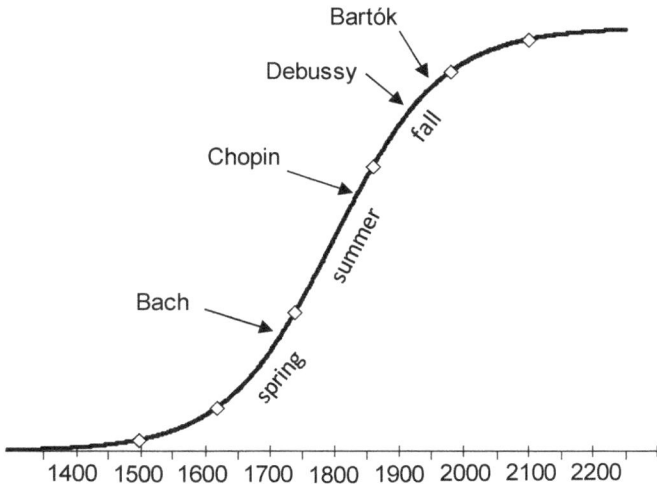

FIGURE 6.7 This S-curve has been constructed by only qualitative arguments and yet it seems accurate and informative. The vertical axis could be something like "importance", "sophistication", or "public's ability to appreciate innovation", (always cumulative). The little diamonds delimit the seasons.

One may wonder why Chopin is more popular than Bartók. Chopin composed during the "summer" of music's S-curve when public appreciation of music grew at a maximum rate. Around that time composers' efforts were rewarded more handsomely than today. The innovations they made in music were assimilated by the public within a short period of time because the curve rose steeply and would rapidly catch up with each innovation. But today the curve has flattened and gifted composers are given very limited space. If they make even only small innovations they find themselves above the curve and there won't be any time in the future when the public will appreciate their work. On the other hand, if they don't innovate, they will not be saying anything new. In either case today's composers will not be credited with an achievement.

7

Cascades of S-Curves

Marketers have long used bell-shaped curves to describe the life cycles of their products. After all, product sales do grow in a competitive environment and the best-fit product normally wins. But marketers have a harder time accepting the fact that natural growth is capped and that their company will not continue growing exponentially, not even linearly. The only way growth can be sustained in a competitive environment is via a succession of S-curve steps as one market niche is filled after the other. Even then, the overall envelope will follow an S-curve pattern on a larger scale.[1]

JUST-IN-TIME REPLACEMENT

A well-known marketers' utopian pursuit is the strategy of product replacement timed so as to avoid any slowdown in the growth of their sales revenue. Evidently, launching products too closely together (the case may be argued for new Microsoft Windows operating systems) may frustrate customers and/or lead to "cannibalization" of their own market when the new product robs sales from the old one. On the other hand, delaying the launching of a replacement product may create a vacuum in a vendor's offerings and result in loss of customers to the competition. So the question becomes when is the optimum time to launch a replacement? The question can be generalized to: when is the right time to introduce change in an ongoing natural-growth process?

No one wants to tamper with something that works well, but how old should become a product before its replacement is launched?

The criterion can be found in harmonic motion and not only because the concept of harmony implies goodness. Regularly spaced product life cycles produce a landscape pattern suggestive of a sine wave, and large-scale growth processes, such as the world energy consumption discussed in Chapter 5, have deviated from a natural trend so as to also produce a sine wave (Kondratieff's cycle). The sine-wave pattern is characteristic of the pendulum's harmonic motion.

I have studied consecutive natural-growth processes by cascading two identical S-curves as a function of the distance in time between them.[2] The curves cascaded in a mutually exclusive way, that is, the first one stopped when the second one took over. A straight line connecting the centers of the two S-curves served as a general trend, see Figure 7.1. The difference between curves and straight line resulted in a cyclical pattern that I fitted with a sine wave. The fit involved the determination of four parameters: the time delay between the two S-curves, and the amplitude, frequency and phase of the sine wave. The fit turned out to be excellent and also very stable against changes in the steepness of the S-curves used.

A Harmonic Cascade of S-Curves

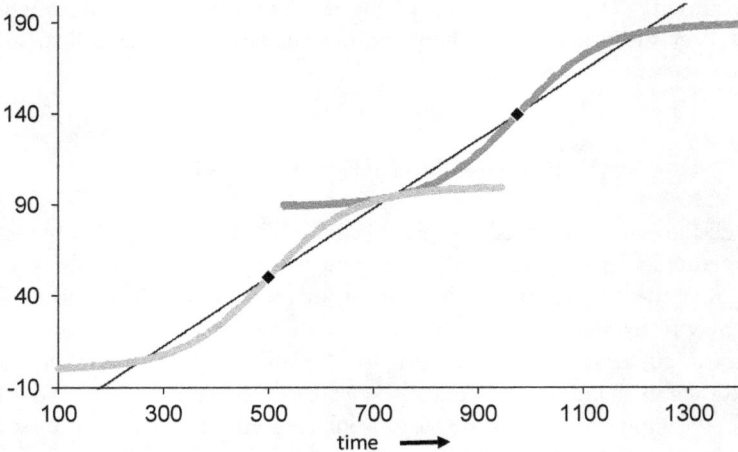

FIGURE 7.1 The same S-curve pattern repeats (dark gray line) from where the previous one (light gray line) leaves off. The straight line connects the center points of the two S-curves. The axes have arbitrary units.

The final overlap of the cascading S-curves is rather small: 2.6%, 10.3%, and 32.7% of the second curve coincide with 89.7%, 97.4, and 99.4 respectively of the first curve, see Figure 7.2.

This recipe spells out in a quantitative way how to harmonically cascade natural-growth processes. It can serve as a quantitative rule for the perennial quest of *just-in-time* replacement. It can be used to time the launching of new products using only the sales information at hand. In order for the new product to achieve 2.6% penetration of its market when the old product reaches 89.7% penetration of its market, the launching of the new product must occur halfway down the phasing-out period of the incumbent product. This point in time falls between the

The Overlap between Two Harmonically Cascading S-Curves

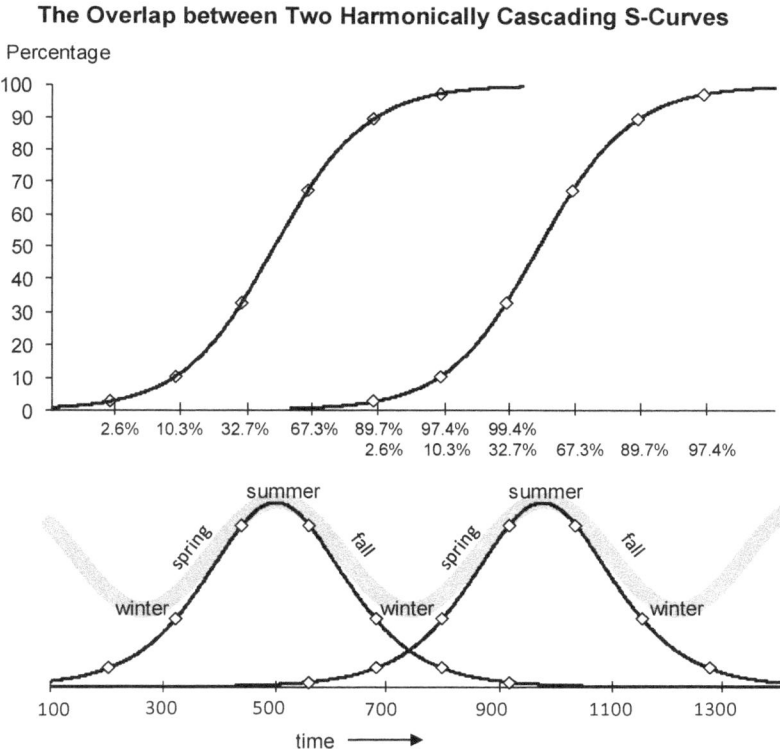

FIGURE 7.2 We see the two S-curves of Figure 7.1 and their corresponding life cycles. The horizontal axis at the middle shows percentage penetration levels for each S-curve. The sum of life cycles cascading in this way matches perfectly the thick gray line, which is a sine wave (harmonic motion). The little diamonds delimit the business seasons. The horizontal axis at the bottom shows the same arbitrary time units as in Figure 7.1.[3]

two little diamonds on the declining side of the life cycle in Figure 7.2, the season of "fall" as defined in Chapter 6. Like farmers sowing in the fall the seeds for the next crop, product managers must sow the seed of their new product during the fall season of the incumbent product.

What we want to retain here is that if natural-growth processes cascade in a harmonic way, then there will be a periodic and regular swing between good and bad business seasons. Harmony in product replacements does not maintain the high growth rates that marketers wish for. The dip in revenue during the replacement phase is natural and inevitable. More than that, it is desirable. This dip plays a significant role in triggering new growth, just as pruning the roses ensures healthier blossoms for the next season. The lesson for business executives facing a major transition between products, technologies, or other fundamental change is not to strive for minimizing its impact but to plan for and anticipate a low-growth period comparable in duration to the high-growth period they just enjoyed.

In personal relationships the winter dip corresponds to a period of mourning. Much has been said by psychologists and couple therapists on the importance of acknowledging the separation, going through a mourning process before moving from one relationship to another.

Proverbial wisdom has long claimed that there is goodness in every season. Some people may think that the most desirable climate is found on tropical islands like Mauritius and Seychelles. Not true! Our above argument based on harmony dictates a large and regular seasonal variation like that encountered in temperate climate, which has also been the cradle of most great civilizations. History is poor in significant cultures that emerged from the arctic or from the tropics, the former perhaps because of conditions hostile to life and the latter mostly because of lack of variation and motivation. Despite an idyllic setting, tropical islands are rather sterile. The expulsion of Adam and Eve from Paradise may have been, after all, an original blessing.

FRACTALS OF S-CURVES

The big picture is not just the cascading from one S-shaped step to the next. It is a cascade of cascades, a stream of S-curves alternating with periods of chaos, as described earlier. Would the corresponding life cycles—the rate of growth—be a regular oscillation? Does the Kondratieff's cycle discussed in Chapter 5 tick like a regular clock over the centuries, repeating its peaks and troughs at exact intervals? Are the random fluctuations of a few percent in magnitude observed during the last two hundred years the only deviations possible? If that is the case, how does one explain the shortening of life cycles (of products, technologies, etc.), and the paradigm shift, progressively crowding humanity's significant transitions in recent history?[4]

A group of people linked together with a common goal, interest, ability or affiliation—for example, a company or an organization—will have its own S-shaped evolution. The individuals now play the role of the cells in a multi-cellular organism, and their assembly becomes the organism. The population in its entirety—that is, the organization itself—can be seen as an individual. The same thing is true for products. A group of products, each with its own life cycle, can be described by a similar but larger curve representing the family or the technology life cycle. If we zoom back further, we may find that many technologies come and go, succeeding each other the way products do. Thus we find a fractal aspect in the S-shaped natural growth pattern.[5] That is, zooming in or out, we obtain the same S-shaped pattern, and the only difference is the timeframe, see Figure 7.3.

The Fractal Nature of Natural Growth

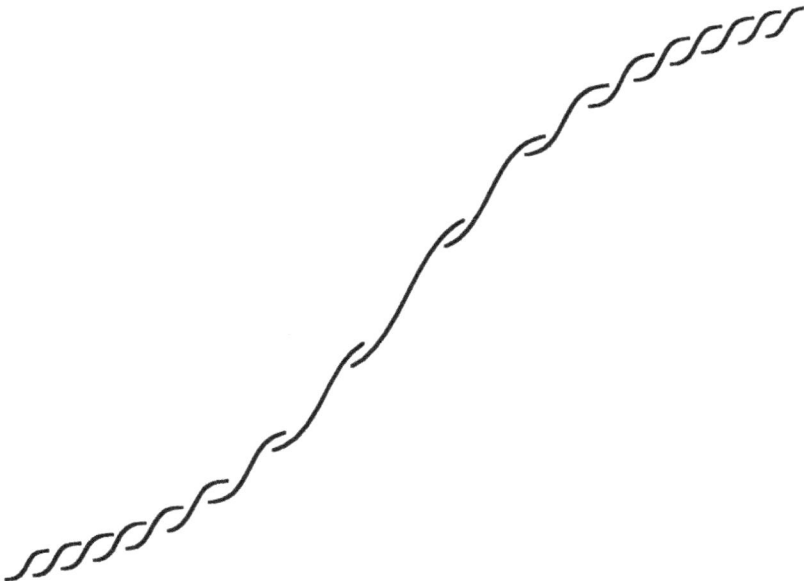

FIGURE 7.3   An overall S-shaped pattern decomposed into constituent S-shaped curves according to a rigorous procedure. The horizontal axis represents time.

S-curves are nested like Russian dolls. Consequently, there are many product seasons in one company season, many company seasons in one industry season, and many industry seasons in one global economy season. In all cases, the main characteristics of the social phenomena and human behavior associated with a given season are similar. But there are non-obvious connections across the different levels of S-curves. For the sake of the discussion, let us consider four levels of nested S-curves. Each level goes through natural growth steps with a life cycle of the same pattern but of different time duration. A typical situation might be as follows:

Products, consist of units sold and may have a life cycle of 6 quarters.

Product families or *companies*, consist of a set of related products and may have a typical business cycle of 5 years—in the case of a product family, 5 years may be the duration of its life cycle.

Basic technologies or *industries*, consist of a set of related product families or companies and can have typical cycles of 15 years.

The economy, is the sum of all industries and has a cycle of 56 years (Kondratieff cycle).

S-curves at different levels are linked. For example, the high output of a summer season feeds the S-curves of one level below. That is why product replacement is least painful when the product family is in summer, just as the replacement of product families is least painful when the technology curve is in summer. Another link between S-curves of different levels occurs with innovation, which normally belongs in winter. Innovation finds itself in the spring season of the S-curves one level above. For example, the spring of an industry brings product innovation. The spring of the economy brings industry innovation.

The graph in Figure 7.3 has been constructed in a rigorous and quantitative way. The overall curve is given a certain thickness within which the constituent S-curves must be contained. Redundant solutions are eliminated by requiring all constituent curves to belong to the same class; that is, they must all have the same slope at their midpoint, so that the maximum rate of growth is the same for all sub-processes. Such a rule has a defensible interpretation in product sales, because it reflects the stability of consumer spending. People's average income (linked to the gross national product) does not change rapidly with time; therefore the buying power of individuals of a certain income class also remains roughly stable over a period of time. As products come and go, rapidly substituting each other, the maximum rate of product sales does not change appreciably on the average, and the larger the product niche, the longer its life cycle. As a consequence, life cycles become longer during the high-growth period and shorter during the low-growth period, as depicted in Figure 7.4.

The large-scale S-curve that becomes visible when we zoom backward serves as an envelope for the succession of the smaller constituent curves. Its level defines the ceiling up to which constituents will grow and its steepness determines the length of their life cycle. The phenomenon of shrinking life cycles, an important concern of manufacturers, can be quantitatively linked to the saturation of the enveloping process, see Table 7.1. For a family of products, shrinking life cycles reflect how close to exhaustion a technology may be. On a larger scale, shrinking life cycles of families of products (for example, mainframe computers) reflect changing social patterns (such as a trend toward demand for more portable and personalized products). On an even larger scale, a large number of families of products—technologies, markets, and so on—with shrinking life cycles may reflect a global economic recession.

Because the decomposition of the envelope S-curve to the constituent product S curves has been done quantitatively, we can relate the shortening of product life cycles to the overall level of saturation— how close it is to exhaustion—of the envelope. We can then monitor the drift of the width of life cycles over time in order to determine either how close we are to full saturation (life cycles getting shorter) or how far we are from a future maximal rate of growth (life cycles getting longer).

In practice, due to technical or other reasons, such as pent-up demand, we often find that the first few short-lived S-curves in a long chain are either understated or missing.

Being able to estimate the level of overall saturation from observing life-cycle trends is a powerful approach. It implies that a non-specialist, such as a dock worker loading boxes onto trucks, may notice that the labels on the boxes change three times as frequently as they used to back in the good old days, and boast to his fellow workers that he knows that the technology behind these products is more than 87% exhausted. He may go further and argue that if things have been done right, the next-technology products should be showing up at the dock with the coming shipment.

The above image may sound naive, but the approach offers valuable insights for tracking of the overall life cycle of products, companies,

The Shortening of Life Cycles

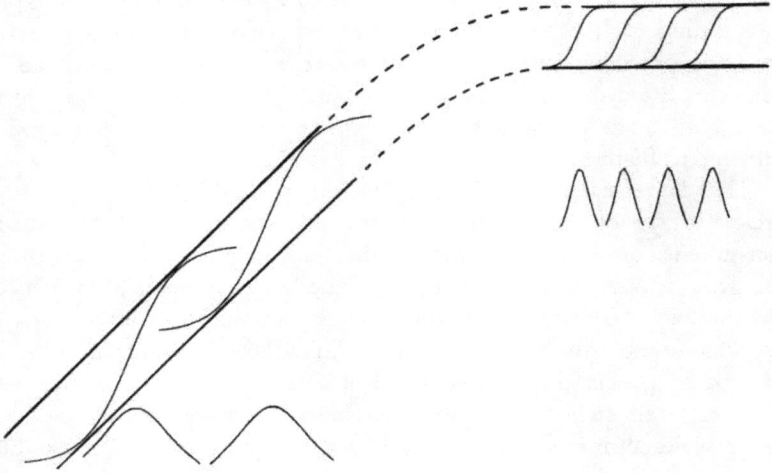

FIGURE 7.4 A geometric explanation of why the end of growth implies shorter life cycles.

Table 7.1 The relation between the shortening of life cycles and saturation

Life-cycle length (relative to longest)	Level of saturation (percent of ceiling)
0.17	3.1
0.19	4.0
0.20	5.2
0.22	6.9
0.24	9.1
0.30	12.8
0.41	20.0
0.70	31.4
1.00	50.0
0.70	68.6
0.41	80.0
0.30	87.2
0.24	90.9
0.22	93.1
0.20	94.8
0.19	96.0
0.17	96.9

technologies, and social trends. Examining the evolution of the sub-processes can give us information about the remaining growth potential of the outer envelope. In other words, the shortening of product life cycles tells about the remaining growth potential of the company. The shortening of company life cycles tells about the remaining growth potential of the industry, and the life cycles of the industries about the whole economy.

An example of nested S-curves can be found in the aviation industry. Wide-body aircraft constitute a family with about a dozen members, each having its own life cycle. Early members, such as the DC10 and Lockheed Tristar, were shorter-lived than the Boeing 747. However, the later rapid appearance of the 767s, a number of Airbuses, MD11s, and 777s implied that these aircraft would have shorter life cycles than the 747s. As in the pattern of Figure 7.1, the wide-body family of aircraft underwent successively the stages of: two short life cycles, one long, and again a number of short ones. We can thus conclude that the overall S-curve, describing the growth process of the wide-body family, is approaching a ceiling, with the 747 as the central long-lived product. In the future, we should expect little—if any—growth in the annual passenger-mile totals of wide-body aircraft. In fact, the average size of airliners on transatlantic flights has already shown signs of decline during the mid-1990s. In that light, the Airbus superjumbo (A380) will not have its own market. This aircraft must steal market share from the wide bodies in use. Even then their sales would probably never reach the volume of 1400 units that Airbus managers had originally anticipated.

We can zoom back and look at all of jet aviation as one family with two members. The first one—early jets—underwent a 15-year growth process. The second one—wide bodies—underwent a 30-year growth process. The picture suggests that there should be a new upcoming type of aircraft—possibly supersonic—with relatively high carrying capacity but narrower fuselage (single corridors) than today's wide bodies. The Concord could be this family's precursor.

Because the life cycles of jet aircraft families have so far been increasing, we conclude that the overall diffusion in jet aviation has not yet passed the midpoint of its S-curve. An independent estimate corroborates this conclusion by positioning the midpoint of this diffusion process in early 21st century.[6] One can thus safely surmise that this new family of supersonic aircraft will grow for longer than thirty years and will constitute the central long-lived family in jet aviation.

THE MICROVAX FAMILY OF DEC MINICOMPUTERS

Digital Equipment Corporation (DEC) became successful thanks to the minicomputer, a market niche created and filled by DEC. The first technology of minicomputers, called PDP, was launched in 1959. In 1978 DEC launched a second architecture of minicomputers, called VAX, based on a 32-bit microprocessor. They replaced half the PDP sales by the mid 1980s. By 1990 DEC launched yet another architecture of computers, called Alpha, based on a 64-bit microprocessor. They replaced half the VAX sales by the mid 1990s. Each architecture had many product families, and each family many products.

As a strategy consultant at DEC, I was asked to study one family of products in detail, the MicroVAXes.[7] They occupied the $20,000 to $50,000 price range. I produced S-curve fits for each one of the computer models in this family and was able to deduce a life cycle for each. The family's first entrant was MicroVAX I. It had a short life cycle, and it was generally considered an unsuccessful product—unfairly so. It should have been seen as a precursor (see our discussion in Chapter 5), an exploratory attempt that paved the way for the products that followed.

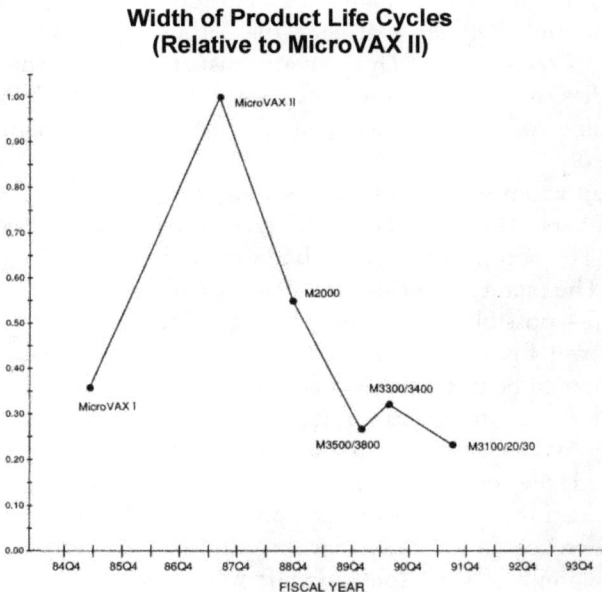

**Width of Product Life Cycles
(Relative to MicroVAX II)**

FIGURE 7.5 The rise and fall of product life cycles in the MicroVAX family of Digital Equipment computers.

MicroVAX II had the longest life cycle—three times as long as its predecessor. The life-cycle duration was already decreasing with the follow-up product, M2000, which lasted for a little more than half as long as MicroVAX II. Later models appeared in rapid succession and featured life cycles more than four times shorter than that of MicroVAX II, see Figure 7.5. According to Table 7.1, this product family should have been around 90% exhausted at the time of my study.

I was able to obtain confirmation of this conclusion by analyzing the growth of MicroVAX as a single process. Despite a limited resolution (only a small number of models), the evolution of MicroVAX corroborated the fractal aspect of logistic growth.

Because of the trend toward smaller computers, the importance of MicroVAX over time was best represented by its market share in the micro-niche considered. The share rose and declined over the lifetime of this product line. Cumulated market share was fitted to an S-curve in Figure 7.6.

The MicroVAX Growth Process
(Normalized Accumulation of Market Share)

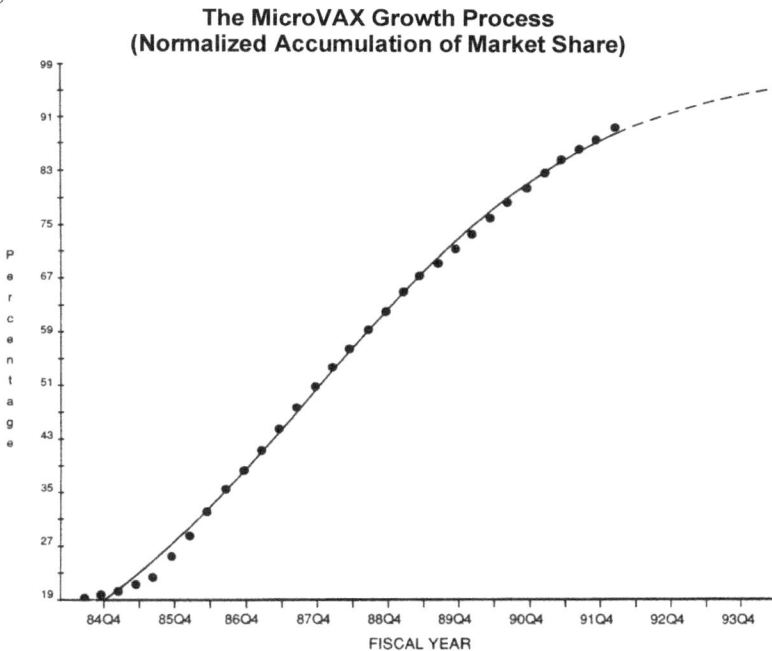

FIGURE 7.6 The MicroVAX story as reflected by the technology's ability to fill its own market niche. The data points represent the accumulation of market share. The line is an S-curve fit up to the second quarter of fiscal year 1992. The process is normalized to 100% at the ceiling of the fitted line. A jump from 0 to 19% at the beginning corroborates the notion of a sudden release of pent-up demand due to a delayed appearance.

In order to focus on the rate of penetration of the MicroVAX technology, the vertical scale was normalized to 100% when the process accumulating market share reached a ceiling. We can thus read saturation levels directly as a function of time. The starting point is 19% reflecting the release of a certain pent-up demand. At the time of my analysis the curve's penetration level was 87% in good agreement with our previous conclusion based on the shortening of the life cycles of this family's component products. It came as no surprise that soon afterward this price range of computers was taken over by workstations and servers.

Following this study, I ventured further on my own knowing that the VAX architecture had many product families like MicroVAX. Zooming back, I was able to see the PDP, VAX, and Alpha technologies all behaving like single products inside DEC's overall company curve. In fact, the life cycles of these architectures were getting shorter. With the life cycle of PDPs taken as unity, the life cycle of VAXes turned out to be equal to 0.65, which, according to Table 7.1, corresponded to 70% completion for the company curve. It also implied that the Alpha architecture would have a life cycle equal to less than 0.4, and therefore should be 50% replaced, by something else, around the year 2002, by which time DEC as a company should be at around the 85-percent completion level of its overall curve.

In fact, DEC never got that far. In June 1998 it was acquired by Compaq, which subsequently merged with Hewlett-Packard in May 2002. Once again, the death of a species came not far from 90-percent completion of its S-curve.

MICROSOFT WINDOWS OPERATING SYSTEMS

The state of maturity of the Windows operating systems as a family became the object of my study when I was asked to review claims for runway trends in the Information Technology industry.[8] Microsoft has been regularly releasing new operating systems. New software developments triggered by hardware improvements make it necessary to introduce major changes to the operating systems. This is a typical characteristic of "young" industries such as microchip production; they are mutational. But as the industry matures, "mutations" become rare and life cycles become longer. Figure 7.7 shows the evolution of the life cycle of Microsoft operating system as defined by the time between release announcements.

Time between Operating Systems

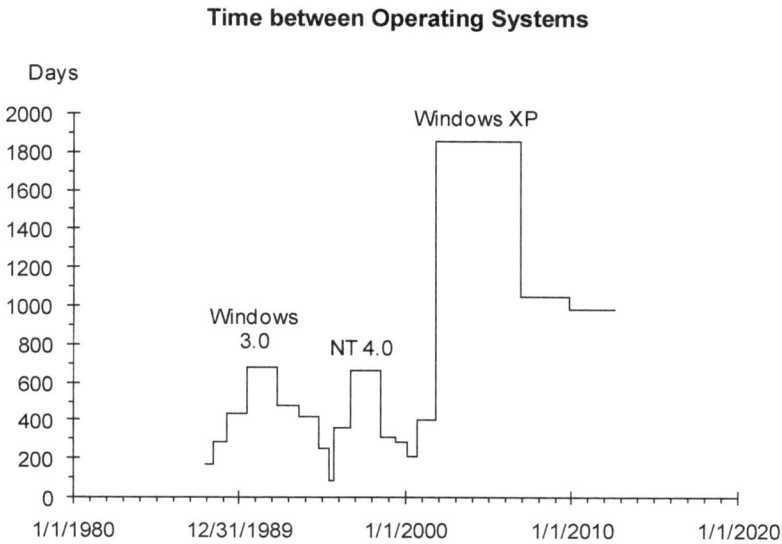

FIGURE 7.7 The duration of Microsoft Windows operating systems as defined by the time between new releases.

Windows XP was the operating system with the longest life cycle. Vista's life cycle was 57% of that of Windows XP and the life cycle of Windows 7 was 53%. With Windows XP at the center of the envelope S-curve for Windows operating systems and the life cycle being symmetric to a fist approximation, we can reasonably expect an end for this growth process by the late 2020s. Appropriately this coincides with an estimate for the end of Moore's law.[9]

Here we were able to make forecasts for the end of Microsoft's Windows operating system as a software architecture from observations on the shortening of the life cycles of individual members of this family. Such forecasts would have been difficult to do—if not impossible—and certainly less defensible using other methods.

WHEN DO WE STOP BEING CHILDREN?

There was an error in my first book *Predictions*. I had written that a newborn's size is around two feet six inches whereas in reality it is closer to twenty inches. The source of my error was that I had backcasted an S-curve fitted on my daughter's height measurements in elementary school and did not bother to go back and check with the original records. Needless to say that it became a source of embarrassment for me when a reader (working in a statistics office) wrote a tongue-in-cheek letter to the publisher wondering whether we were all bachelors!

On my part I ran to the medical library and obtained full sets of data on the evolution of the human size from conception to old age. It became immediately obvious that one S-curve could not describe the entire process. I tried to describe the complete dataset with as few S-curves as possible. I was able to discern three different processes, see Figure 7.8.

The first S-curve covers the period of gestation and extends to sometime beyond the age of one. The second S-curve covers from two to twelve years of age. The excess between ages thirteen and nineteen was amenable to a third S-curve fit. The goodness of these fits deteriorated rapidly with changes in the limits of the corresponding the datasets.

The Stages of Human Growth

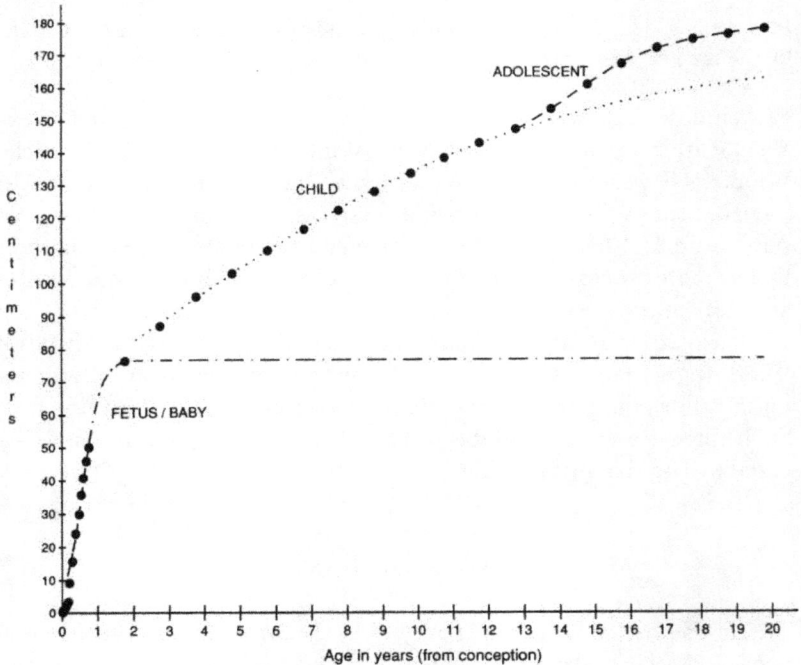

FIGURE 7.8 Data and three S-curve fits for the phases of growth in size of a human male.[10]

All three S-curves are partial, so all three fits involved a fourth parameter, a pedestal (positive or negative) on which the S-curve is sitting. For the middle curve, corresponding to childhood, we see only half of an S-curve, but it contributes the largest-size chunk to one's height and covers more than 90% of one's total-growth period. Second in importance is the fetus/baby growth process with a comparable contribution to the final height but a much shorter time constant. Adolescence is a higher-order effect (is less pronounced for girls) that begins at age thirteen and contributes less than 10% of the total height. A more in-depth study may discern finer logistic-growth processes around the egg and the embryo states. Finally, there is a relatively minor downward pointing S-curve describing the shrinking of one's height with old age, not shown in the figure.

My study of the human-size growth did not only serve to appease my publishers and reply to the vocal critic of my book. It also produced a number of insights. Evidently everyday terms such as fetus, baby, child, and adolescent have precise definitions reflecting growth processes that are distinct. Also the fact that fetus and baby belong to the same growth curve must be related to the fact that the distinction between these two states is largely "geographical", meaning inside or outside the womb. After all, the transition between them takes place abruptly, the ratio of the durations between these two stages may vary as much as a factor of 3 from case to case, and the exact time of transition is most often arbitrary (induced labor and/or caesareans).

IN CLOSING

Sustained growth consists of successive S-shaped steps, each of which represents a well-defined amount of growth. A new S-curve begins where the previous one left off. Every step is associated with a niche that opened following some fundamental change (a mutation, a major innovation, a technological break-through, etc.). But all transitions display the characteristics of winter seasons.

When 10.3% of the replacement process overlaps with 97.4% of the incumbent we have a harmonic cascade of natural-growth processes. The everyday words harmonic and natural here have rigorous science-based definitions. The former concerns motion under a restoring force that is proportional to the displacement (harmonic motion) and the latter concerns a rate of growth that is proportional to the amount of growth already achieved but also to the amount of growth remaining to be achieved (natural growth in competition).

Between successive growth phases there is a lull in the overall rate of growth, a winter season that is both unavoidable and beneficial. Just-in-time replacement of products does not maintain constant or increasing sales. It ensures that a dip in overall sales will be of natural duration and size. Its duration should be comparable to the duration of the rapid-growth phase experienced while the last S-curve was steeply rising. The swing between high-growth and low-growth seasons stimulates the organism's vitality, and the timely anticipation of the winter season allows for the appropriate strategic planning that will contribute to the organism's longevity.

An old mathematical pastime has been describing *any* function as a sum of terms belonging to a class of functions, for example the Fourier analysis that decomposes functions into sines and cosines. The most slowly varying component—responsible for the main trend—is referred to as the *fundamental harmonic*.

In this tradition one may want to decompose *any* overall growth process into S-curve sections instead of sines and cosines and for a given process we may find that one S-curve component plays a major role, even if the overall pattern shows significant deviations from a large-scale S-curve. In an analogy with Fourier analysis we would say that this is the fundamental harmonic and that higher harmonics are expected to play less important roles. For the case of human-size growth discussed earlier the "fundamental harmonic" would be childhood.

8

Genetic Reengineering of Corporations

— How can our role in the market become more predator and less prey?
— Should we go for differentiation or for counterattack?
— Is our advertising budget spent in the most productive way?
— Do we need to change our image?

• • •

These are questions that managers would love to have answered in a trustworthy way. But most of the time they muddle through these decisions without really knowing what they are doing. A model elaborated on S-curves (known as Volterrra-Lotka equations) can give science-based answers to such questions. Let us see how this is done.

An S-curve describes the growth in competition of a species population. Competition results because members of the same species elbow one another in a crowded niche. In the presence of more than one species, however, the S-curve law does not generally apply, because one species may interfere with the growth rate of another in many ways. More terms must be added to the mathematical formulation to take this into account, and the S-shape pattern becomes distorted. An exception is one-to-one substitutions: They involve two competitors, and yet their "market" shares follow S-shaped patterns (see the case of cars and horses discussed in Chapter 3).

There are two bends in the graceful shape of the celebrated S-curve. The first one (exponential rise) is due to the capability of the species to multiply. The second one (niche-saturation slowdown) is due to the competitive squeeze caused by the limited space.

The first bend. Initially products sell into a market niche just like rabbits fill an ecological niche, namely exponentially. If the average rabbit litter is taken as 2 you can watch the rabbit population increase by the successive stages of 2, 4, 8, 16, 32,..., 2^n, in an exponential growth. If the average rabbit litter is greater than 2, you will see a steeper exponential growth. The same is true with products. Depending upon its attractiveness—the equivalent of the average rabbit litter—every product sold will bring new customers. The more products out there and the more attractive they are, the higher the rate of sales. Sales will grow at a constant percent rate—that is, exponentially—for a while, with a time constant defined by the product's attractiveness. (If a product's attractiveness is smaller than unity, each product sold will bring less than one new customer, indicating that we are dealing with an unsuccessful entry, and its sales will quickly dwindle down to zero.)

The second bend. The rabbit population explosion ceases when a sizable part of the niche becomes occupied; the same is true with products. The growth equation that is valid for late *as well as* early times contains a second factor: the fact that the niche capacity is finite. In other words, the equation says that the percentage rate of growth is proportional not only to the product attractiveness but also to the still-empty space in the market niche (see Appendix B).

MORE THAN ONE SPECIES IN THE SAME NICHE

Two parameters, the attractiveness and the niche capacity, fully determine the S-shaped pattern evidenced in the evolution of a species population diffusing in its ecological niche. But what happens if besides rabbits we also have sheep on the range? After all, sheep also eat grass and in greater amounts than rabbits. Their presence will certainly suppress the rabbit population explosion. Worse yet, what happens if there are foxes? Competition between rabbits and sheep is not the same as between rabbits and foxes. Just think of the fact that, faced with a finite amount of grass, sheep would probably lament at the rapid multiplication of rabbits, whereas foxes would undoubtedly rejoice.

The basic mechanism is how one competitor influences the growth rate of the other. Sheep and rabbits have a negative effect on each other's population by reducing each other's food supply. In contrast, foxes damage rabbit populations, while rabbits enhance fox populations. Whenever there is more than one competitor in the same niche, we must consider the interaction between them— namely, how one's rate of growth depends on the existence of the other. We then need to introduce a third parameter in the growth equation to take this coupling into account. The value of this parameter is related to the overlap between competitors, or how much one steps on the other's feet—in other words, how many sales you will lose (or win) because your competitor won one. This way we can formulate a measurement for our ability to attack, counterattack, or retreat, as the case may be.

ATTACKER'S ADVANTAGE, DEFENDER'S COUNTERATTACK

The attack of a new species against the defenses of an incumbent lies at the heart of corporate marketing strategies. This kind of struggle had already been rigorously formulated by biologists and ecologists. In the 1930s George Gause, at Moscow's Zoological Museum, studied the competition between traditional brewer's yeast and one used in Ukraine to make the refreshing milk drink called kefir, popular in Asian and Middle-Eastern countries. He first grew the two yeasts in isolation and observed the S-shaped natural growth pattern for each. He then put them together in the same test tube and let them compete for the same food. He found that each influenced the other's growth. But the brewer's yeast is tolerant to the alcohol that is produced as it grows; the kefir yeast is less so. In a mixture, the brewer's yeast thus had an increasing advantage as fermentation proceeded, and it outgrew its competitor. Simple S-shaped curves did not describe the growth processes well because they could not handle the interference of the growth rate of one on the other. But the Volterra-Lotka mathematical formulation involving coupling constants can do that.

Christopher Farrell, director of scientific affairs at Baxter Healthcare, defined an *attacker's advantage*, and a *defender's counterattack* in terms of the coupling parameters in the growth equations.[1] The attacker's advantage quantifies the extent to which the attacker inhibits the ability of the defender to keep market share. The defender's counterattack quantifies the extent to which the defender can prevent the attacker from stealing market share. The business strategy and tactics of attack and counterattack have been qualitatively described by Peter Drucker,[2] and especially by Richard Foster, director at McKinsey & Co.[3] The nature of the attacker's advantage has been clearly established by Cooper and Kleinschmidt (1990)—professors respectively in industrial marketing and technology management, and in marketing and international business—who studied over 200 new products and determined that the most significant parameter in gaining market share is a "superior product that delivered unique benefits to the user."[4] This and price considerations dictate the magnitude of the attacker's advantage.

Under attack, the defender redoubles its own efforts to maintain or improve its position. A high value for the defender's counterattack implies a face-on counterattack within the context "we do better what they do". An effective counterattack, however, with long-lasting survival-sustaining consequences implies eventual adoption of the new technology, some sort of death for the old company, an end that is painful to assimilate culturally. Because companies hesitate to embark on such undertakings, Foster refers to defender's counterattack as the defender's *dilemma* and cites tens of examples in which a defender refused to acknowledge, or reacted too late, to an attacker's onslaught. A classical case was NCR's belated and traumatic transition to computerized cash registers.

Figure 8.1 shows the early days of the competition in the Greek mobile-telephone market, a two-competitor struggle. Panafon and Telestet launched their products simultaneously. One firm, Telestet, became an early market leader, thus assuming the role of the defender. But the coupling parameters, determined from the data, were both negative and significant, the attacker's advantage = -0.8 and the defender's counterattack = -0.6. The figures indicated much overlap and fierce competition of the sheep-rabbit nature. Every time Panafon would close a sale, Telestet would lose 0.8 potential sales, and every time Telestet would close a sale, Panafon would lose 0.6 sales. The difference was crucial. The model showed curves that eventually deviated from S-shapes. With data up to the end of 1995, the model's prediction was that Panafon would become leader within a few months. By mid 1996 Panafon's market share had been established higher than Telestet's.

Kristina Smitalova and Stefan Sujan—professors of mathematics at Comenius University and the Slovac Academy of Science respectively, in Bratislava, Slovakia—studied and classified the various coupling schemes in a rigorously.[5] They distinguished and labeled six ways in which two competitors can influence each other's growth rate, according to the sign of the two coupling parameters involved. The tabulations are shown in the boxed text.

Mobile-Telephone Sales in Greece

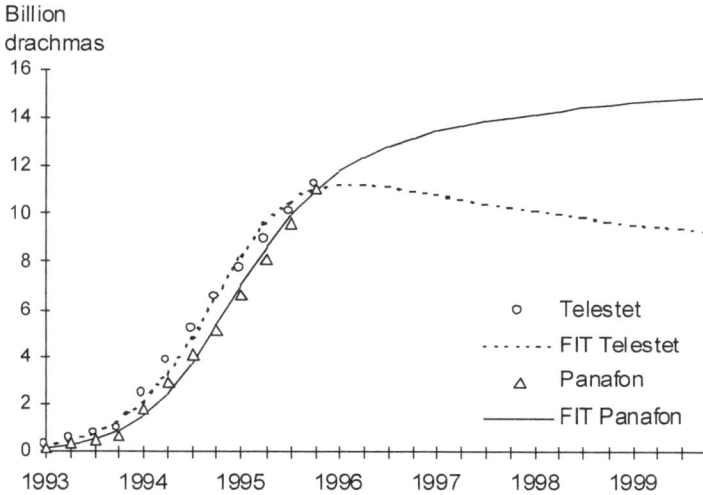

FIGURE 8.1 Quarterly sales for the two competitors of the Greek mobile-telephone market. Despite early dominance by Telestet, the model successfully predicted the shift in Panafon's favor by mid-1996.

– –	*Pure competition* occurs when both species suffer from each other's existence.
+ –	*Predator-prey* occurs when one of them serves as direct food to the other.
+ +	*Mutualism* occurs in case of symbiosis, or a win-win situation.
+ 0	*Commensalism* occurs in a parasitic type of relationship in which one benefits from the existence of the other, which nevertheless remains unaffected.
– 0	*Amensalism* occurs when one suffers from the existence of the other, which is impervious to what is happening.
0 0	*Neutralism* occurs if there is no interaction whatsoever.

Pure competition is what we have between rabbits and sheep. Each one diminishes the growth of the other but not with the same importance (sheep are fewer but eat more). Market examples are the mobile-telephone case mentioned above and the competition between different-size computer models.

Predator-prey is the case of cinema and television. The more movies made for cinema, the more television will benefit, but the more television grows in importance, the more cinema suffers. Films made for TV are not shown in movie theaters. Had been no legal protection (restricting permission to broadcast new movies), television would have probably "eaten up" the cinema audience.

A typical case of mutualism is software and hardware. Sales of each trigger more sales for the other.

Add-ons and accessories such as car extras illustrate commensalism. The more cars sold, the more car accessories will be sold. The inverse is not true, however; sales of accessories do not trigger car sales.

Amensalism can be found with ballpoint pens and fountain pens, described in detail below. The onslaught of ballpoint sales seriously damaged fountain pen sales, and yet the ballpoint-pen population grew as if there were no competition.

Examples of neutralism are encountered in all situations in which there is no market overlap—for example, a sports store that sells both swimming wear and ski wear. Depending on the geography there might be a negative correlation of seasonal origin, but the sales of one product do not in general affect the sales of the other.

COMPETITION MANAGEMENT

The intriguing fascination of the marketplace is that *the nature* of competition can be changed over time. For some businesspeople, achieving a change in the competitive roles is perhaps more handsomely rewarding than making profits. It is something that species in nature cannot do. Rabbits will never eat meat, and whenever humans tamper in such areas, either academically (genetic engineering) or industrially (mad cow disease), they are invariably criticized, justly or unjustly.

But things are different in industry. In contrast to the jungle, a technology, a company, or a product does not need to remain prey to another forever. The competitive roles can be radically altered with the right decisions at the right time. External light meters, used for accurate diaphragm and speed setting on photographic cameras, enjoyed a stable *commensal* relationship with cameras for decades. As camera sales grew, so did the sales of light meters. But eventually technological developments enabled cameras to incorporate light meters into their own box. Soon the whole light-meter industry became prey to the camera industry. Sales of external light meters diminished while sales of cameras enjoyed a boost, and the relationship passed from *commensalism* to a *predator-prey*.

The struggle between fountain pens and ballpoint pens, mentioned earlier, had a happier ending. Another case of genetic reengineering in the marketplace, the substitution of ballpoint pens for fountain pens as writing instruments went through three distinct stages.

Before the appearance of ballpoint pens, fountain-pen sales were growing undisturbed to fill the writing instrument market. They were following an S-shaped "rabbit curve" when the ballpoint technology made its appearance in 1951. As ballpoint sales picked up, those of fountain pens declined for the period 1951-1973. Ballpoint pens did not belong to the same species or even constitute a one-to-one substitution, and yet they cut deep into the fountain pen sales. A simple S-shaped pattern could not have described this transition, but the Volterra-Lotka equations did, with attacker's advantage = -0.5 and defender's counterattack = 0 (see Figure 8.2).[6] These numbers imply a competitive

The Struggle between Ballpoint and Fountain Pens

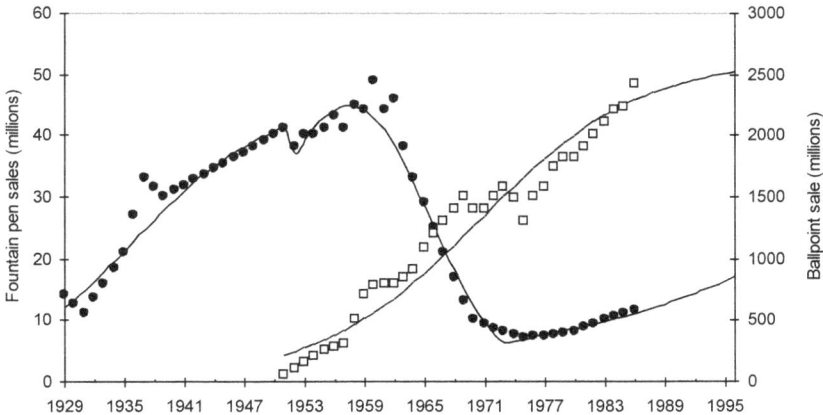

FIGURE 6.2 Sales of fountain pens and ballpoint pens in the US. The lines are our model's descriptions. Before 1951 and after 1973 we see *plain* S-shaped patterns for each competitor. Between 1951 and 1974 we see a typical *amensal* type of competition, in which the attacker has an advantage (attacker's advantage = - 0.5), and the defender's counterattack is null (defender's counterattack = 0).

advantage for ballpoint pens, which by winning one customer inflict losses of half a customer to fountain pens. Fountain pens staged a counterattack by radically dropping prices for many years. Their average price dropped as low as 72 cents. But the counterattack was ineffective—*defender's counterattack* remained equal to zero. Counterattacking fountain pens lost market share and embarked on a well-established extinction course.

Eventually the prices of fountain pens began rising. The average pen price in the United States reached $3.50 in 1980 and continued rising. In 1988 a Mont Blanc Masterpiece Diplomat retailed at $280, while a Waterman Le Mans 100 Briarwood cost $400. The fountain pen underwent what Darwin would have described as a "character displacement" to the luxury niche of the executive pen. The strategy of fountain pens since the early 1970s has been a retreat into noncompetition. Indeed, the *attacker's advantage* and the *defender's counterattack* must both equal to zero for the Volterra-Lotka equations to do justice to the sales data of writing instruments in this period. In other words, we have two species that do not interact—*neutralism*—but each follows a simple S-shaped growth pattern. As a consequence fountain pens have secured for themselves a healthy and profitable market niche. Had they persisted in their competition with ballpoint pens, they would have perished.

Now that we have quantified the competitive mechanisms from 1951-1973, it is amusing to play the following scenario: What would have happened if fountain pens had undergone their character displacement five years earlier? The model's answer is a significantly higher number of sales for fountain pens today. Is it believable?

Arguably so. Fountain pens would have embarked on an upward trajectory earlier, starting from a stronger position. Enhanced fountain pen content in everyday life could have cultural repercussions over time, producing societal preferences and habits. In the end, a more favorably disposed average citizen could have meant a more important role for fountain pens today. Consequently on the average, their price would have had to rise less and their image would be a little more popular and a little less exclusive.

Character displacement is a classical way to diminish the impact of competition. Another name for this is Darwinian *divergence*, encountered among siblings. In his book *Born to Rebel*, Frank Sulloway—a historian-of-science turned sociologist at MIT's program in science, technology, and society—proves that throughout history first-born children have become conservative and later-borns revolutionary. First-born children end up conservative because they do not want to lose any of the only-child privileges they enjoy. But this forces later-borns into becoming rebellious, to differentiate themselves and thus minimize competition and optimize survival in the same family.[7]

FINDING THE MAGIC ADVERTISING MESSAGE

The Volterra-Lotka equations require three parameters per competitor to describe growth in a two-competitor niche. One parameter represents the ability to multiply, another the size of the niche, and the third interference from the other competitor. Consequently, there are three choices for action; or six, if we want to take into consideration the parameters of the competitor. To increase the prospects for growth then, we can try to change one or more of the following:

- The product attractiveness (increase ours or decrease theirs)
- The size of the market-niche (increase ours or decrease theirs)
- The nature of the interaction (increase our attack or decrease their attack)

Each direction of action in principle affects only one parameter.[8] But it is not obvious which change will produce the greater effect; it depends on the particular situation. The concrete actions may include performance improvements, price changes, image transformation, and advertising campaigns. Performance and price concern "our" products only, but advertising via the appropriate message can in principle influence all aspects of competition, producing an effect on all six of the parameters. The question is how much of an effect a certain effort will produce. Some advertising messages have proven significantly more effective than others. Success is not necessarily due to whim, chance, or other after-the-fact explanations based on psychological arguments. The roles and positions of the competitors at a given point in time determine which advertising message will be the effective one, see Figure 8.3. The effectiveness of advertising messages can be illustrated by a typical competitive technological substitution: woven carpets and tufted carpets.

Woven carpets were made on a loom in a manner similar to plain cloth, except that extra wrap yarns were introduced and raised by wires to form loops. Most of today's carpets are made with needles that punch loops through a backing and retreat to leave tufts. Examining the backing of a

The Six Dimensions of Advertising Action

	Attractiveness	Niche size	Competition
WE:	↗ Our products are good	↗ You need our products	↘ We are different
THEY:	↘ Their products are not good	↘ You do not need their products	↗ We do better what they do

FIGURE 8.3 The six possible independent advertising messages according to our model.

typical modern carpet reveals that glue holds the tufts in place. This revolution in carpet making began in the 1950s. Tufting changed the requirements for the yarn. Long, continuous filaments were preferred, as they didn't pill or fuzz. Wool yarns have fibers as short as the annual growth of a sheep's hair. A fiber such as nylon thus moved into very good position, especially when DuPont invented a bulked form of continuous fiber. The combination of bulking and tufting created a new "species" that satisfied a growing demand for carpeting and caused the displacement of woolen woven carpets by nylon tufted ones.[8]

The model description of the data indicates that the *attacker's advantage* = −2.2 and the *defender's counterattack* = −2.6. This is a typical situation of *pure competition* between two similar-species contenders even though the attacker sells in greater numbers. The fate of the defender is eventual extinction.

Could the makers of woolen woven carpets have secured for themselves a market niche the way fountain pens did? If so, what line of action should have they followed? Let us explore alternative lines of action—via advertising campaigns—and their effectiveness in shaping a different future for woolen carpets back in 1979. Let us rate the different scenarios stemming from changing the six parameters one at a time *by the same amount*, which to a first approximation can be taken as equivalent to comparable-effort investment. It is a sensitivity study on the effectiveness of the corresponding advertising message.

Figure 8.4 shows two of the six possible results.[9] Effective campaigns would be those that emphasized attractiveness and differentiation with messages like "Wool is good" and "Wool is different from nylon." However, a counter-attack along the lines: "Wool is better than nylon", would have been very ineffective.

Substituting Nylon-Tufted Carpets for Woolen-Woven Ones

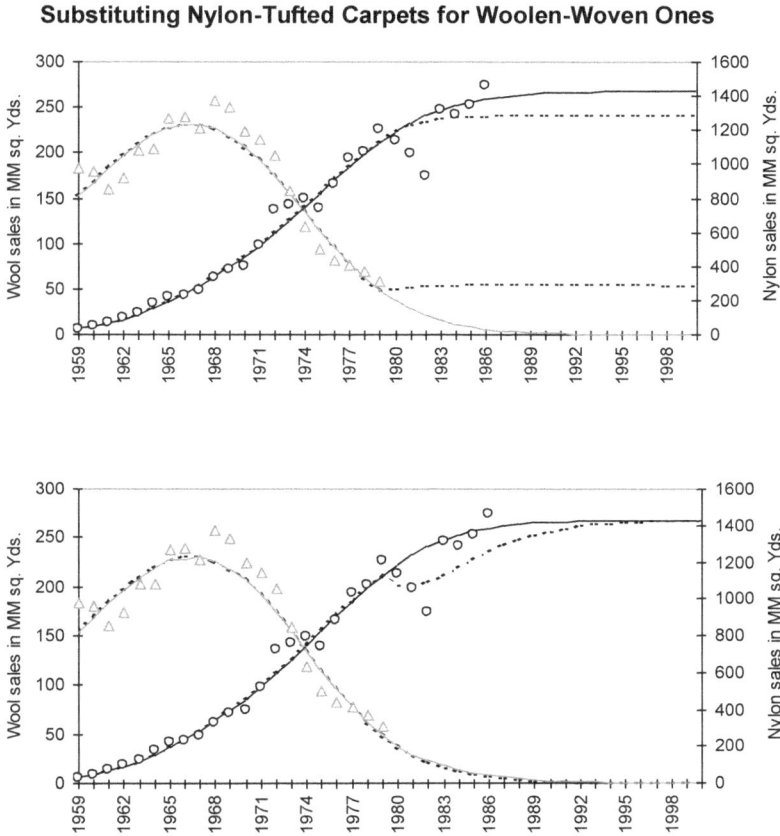

FIGURE 8.4 The dotted lines indicate two scenarios of comparable change in the respective parameters: on the top following an advertising campaign under the slogan "Wool is different from nylon"; on the bottom under the slogan "Wool is better than nylon".

Table 1 shows the complete list of possible advertising messages and their effect on the evolution of wool and nylon sales in 1979. Each message represents an *independent* direction in which the full traditional advertising machinery would have to be launched. There should be no crosstalk between directions. For example, to obtain maximum benefit from the "Wool is good" a campaign *should not* mix connotations such as "Wool is better than nylon" or "Nylon is bad". Each message would have to be developed and exploited separately.

Table 1 Sensitivity Analysis

MESSAGE	EFFECTIVENESS	WOOL	NYLON
Wool is good.	Highest	Slowly rising from '79 level	Little compromised
Wool is different from nylon.	High	Stabilizes at '79 level	Little compromised
You do not need nylon.	Medium	Stabilizes at 0.5 of '79 level	Huge loss of market
Nylon is bad.	Poor	Stabilizes at 0.3 of '79 level	Serious loss of market
Wool is better than nylon.	Negligible	Null	Temporary losses only
You need wool.	Null	No effect	No effect

Although the detailed execution of the advertising campaign (media, wording, style, and so on) remains crucial, the effectiveness ratings of the above directions come in a non-obvious if not surprising order and could not have been arrived at by intuitive or other methods used by advertising agencies. Furthermore, the order may be completely different at another time or in another market.

Table 1 shows the results of the complete sensitivity analysis following six scenarios played for the wool-nylon case study. Playing the scenarios from wool's point of view, we measure effectiveness according to how much wool benefits (see Appendix B for the technical details).

Finally, there is a way to assess the size of the advertising investment called for. An advertising campaign along the lines of "Our product is good" affects the product's attractiveness just as a price cut does. (Price can be quantitatively related to attractiveness via price elasticity). The costs incurred from price dropping alone can thus be compared to those of an advertising campaign that achieved the same result. Naturally, this assessment may result in an overestimate or an underestimate depending on how the advertising campaign in question rated to the "Our product is good" alternative in the sensitivity analysis. It should be noted, however, that if the survival of woolen carpets depended on price dropping alone, the price would have to be cut by more than 100 percent!

The case of the Greek mobile-telephone market described earlier in Figure 8.1, is more malleable. As indicated, Telestet could have anticipated its eventual loss of the leading position. If its managers had taken action in the beginning of 1996 toward increasing the attractiveness of its products by 10 percent—for example, by dropping prices by 8 percent—Teleset would have safeguarded its lead. Of course, Panafon may have rapidly responded in kind, but this is what the business game is all about—and to a large extent it can be successfully, and painlessly, simulated on your personal computer!

NOBEL LAUREATES

We have looked at the number of Nobel laureates won by Americans using S-curves in Chapter 5. The assumption was that there is a limited resource, the total number of Nobel laureates that the US will ever claim. The implication was that this number is capped. In other words, there will be a time when all Nobel Prizes will be awarded to nationals from other countries. Up to that time, Americans will be elbowing each other to win prizes and the fewer left in their "niche" the harder it will be to win one. This reasoning gave rise to an S-curve for US Nobel awards.

But there is also competition between Americans and other nationalities, which was briefly looked at in Figure 3.7. The Volterra-Lotka equations can shed more light in this situation in a rigorous way. To the extent that US Nobel laureates represent about half of all Nobel Prizes every year, it is a good approximation to consider a duopoly, that is, a niche with only two species: Americans and all others grouped together. The species "all others" is rather inhomogeneous but with US Nobel laureates and all Nobel laureates both being well defined as species candidates, "all others" also becomes a well-defined species candidate.

The Volterra-Lotka equations require three parameters per competitor to describe growth in a two-competitor niche. One parameter represents the ability to multiply, another the size of the niche, and the third interference from the other competitor. A total of six parameters must be determined by fitting the Volterra-Lotka equations in a two-competitor niche.

Because there were large fluctuations on the yearly data the numbers were grouped together inside time bins of decades before the analysis. The fit was of acceptable quality (70% confidence level) and the results are graphed in Figure 8.2 and tabulated in Table 8.1. The American trajectory is S-shaped (but not exactly an S-curve) and the long-term forecast is roughly a 50-50 split of all Nobel Prizes between Americans and all other nationalities.[10]

Of particular interest are the values of the coupling constants. They are both positive indicating a win-win nature for this competition. In a symbiotic relationship each competitor benefits from the existence of the other, which is in line with the dynamics of scholarly research (each publication triggers more publications). But Americans benefit more when non-Americans win Nobel Prizes than vice versa. The ratio c_{xy} / c_{yx} is about 1.5 implying that one Nobel Prize won by a non-American will trigger 1.5 times more Nobel Prizes for Americans than the other way around. This is counteracted to some extent by the smaller attractiveness constant for Americans, which reflects the species' ability to multiply. If it is greater than 1, the species population grows; if it is less than 1, it declines. The values in Table 8.5.4 translate to attractiveness for Americans and all others of 1.5 and 1.7 respectively. This indicates that each American Nobel laureate generally "broods" 1.5 new American Nobel winners whereas for all others this number is 1.7.

All in all, the number of American Nobel Laureates in the long run should stabilize around an average of 61.4 per decade barely higher than 60.6 for all others.

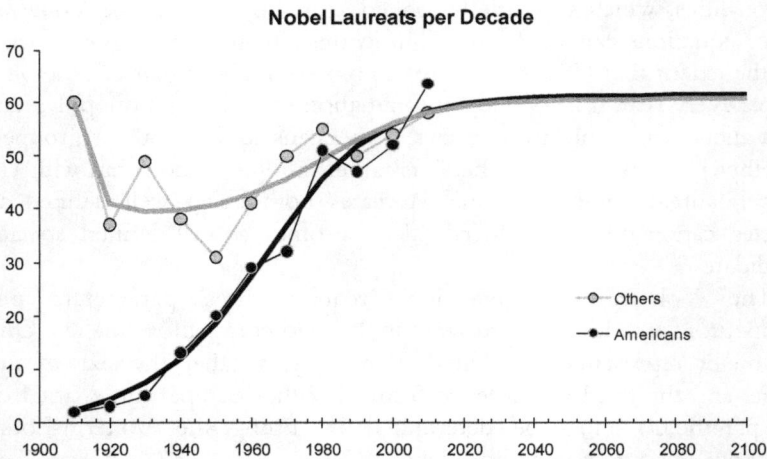

Nobel Laureats per Decade

FIGURE 8.2 Decennial data points and solutions to the Volterra-Lotka equations (the most recent points—awards for year 2010 not known at the time—had been scaled up by 10/9). Despite its S-shaped form the black line is only *approximately* logistic.

All conclusions need to be interpreted within the uncertainties involved. Confidence levels of 70% indicate that there are 7 out of 10 chances that the Volterra-Lotka description is the right way to analyze this competition, which gave results not very different from the S-curve fits in Chapter 5. For the intermediate future—ten to twenty years—the logistic normalized to reasonable population projections would result in forecasts compatible with those of the Volterra-Lotka approach. Still, I would favor Volterra-Lotka because it addresses a more general type of competition. In any case, very long-term forecasts cannot be reliable and the whole exercise must be repeated with updated data sets in a couple of decades, by which time it may be appropriate to consider more than just two players.

Interestingly the US trajectory turned out S-shaped, which suggests that a logistic fit could have been a reasonable approximation but not for the cumulative numbers. The fit should have been on the numbers per unit of time. The limiting resource in this case would have been the annual number of American laureates. This number was zero at the turn of the 20th century and progressively grew to 8 by 2009 (6.4 on the average during 2000-2009). The meaning of competition in this picture would be that Americans elbow each other every year for one of their "quota" prizes that grew along an S-curve and in the 21st century reached a ceiling of 6.1.

The forecasts for American Nobel laureates from the Volterra-Lotka approach are comparable to the number of Nobel laureates won by all other nationalities together. But the fitted parameters gave rise to some interesting insights. The competition between Americans and all others for Nobel Prizes is of the win-win type. Locked in a symbiotic relationship both sides are winning but Americans are profiting more by 50%. At the same time, the ability of Nobel laureates to "multiply", i.e. the extent to which a Nobel laureate incubates more laureates, is lower for Americans than it is for other nationalities. One may ponder whether the roots of this last observation have something to do with the fact that chauvinistic traits tend to be more endemic in cultures with longer traditions.

Table 8.1 The Six Parameters of the Volterra-Lotka Equations

	Attractiveness	Niche size	Competition
Americans	1.5	26	0.6
Others	1.7	37	0.4

WHO IS AFRAID OF THE BIG BAD Wolf?

The S-curve model enhanced with two-species interactions, as presented above, accounts for the three most fundamental factors that shape growth: the attractiveness of an offering, the size of its market niche, and the interaction with the competitor. (When there is more than one competitor, the situation can always be reduced to two players by considering only the major competitor grouping all others together). Naturally, other factors influence growth, such as channels, distribution, market fragmentation, total market growth, market share, frequency of innovations, productivity in the ranks, and organizational and human resource issues. Many factors can be expressed as combinations of the three fundamental ones. Alternatively, the model could be elaborated— by adding more parameters—to take more phenomena into account.

As it stands, the model provides the baseline, the trend on top of which other, higher-order effects will be superimposed. It guides the strategist through effective genetic manipulations of the competitive roles in the marketplace. It should be used as a front end to what is usually done. The model works equally well for products, for corporations, for technologies, or for whole industries. Only the time frames differ. The pleasure is all the strategist's, who now has a quantitative, science-based way to understand the crux of the competitive dynamics and to anticipate the consequences of possible actions.

Just think of it—at this very moment there may be a cost-effective way to terminate the state of being prey to the voracious competitor that has been feeding persistently on your achievements.

9

The Primordial S-Curve and the Singularity

One autumn day in 1999 while my daughter was bragging about her new cell phone's amazing features, I pointed out to her that there were even newer features that it was missing. She said she knew about them, and suddenly turned toward me in desperation, "technology is going too fast!" she exclaimed. I thought I discerned a glimpse of terror in her glance.

I had noticed the accelerating rate of change and not only concerning technology. But my knowledge of S-curves assured me that exponential growth is encountered only during the early stages of natural-growth processes. Eventually natural growth must slow down and therefore the frenzied appearance of change in our lives could not expand forever.

A week later I traveled to America for a consulting project and among other engagements I also got in touch with a physicist friend of old times, who was now professor of cognitive and neural systems at the University of Boston. We hadn't seen each other since we had worked together in Brookhaven National Labs as graduate students doing our theses.

The reunion was elating. We spent hours reminiscing about old times and recounting new experiences. At one point I mentioned my belief that the exponential rate of change should eventually begin slowing down. My friend jumped. Apparently it was a topic he had been preoccupied with. In fact, he had tabulated what he called the twenty most significant turning points in human evolution, such as the discovery of the steam engine and the printing press, the transistor, computers, the Internet, and the sequencing of the human genome. Obviously, his milestones were crowding toward recent times. In fact, he was able to demonstrate that they had followed an increasingly accelerating pattern for 15 billion years!

I was excited. I thought this topic could yield a stimulating publication, but it was imperative to find *objective* data. My friend's twenty milestones could not stand up to the likely criticism of having a biased dataset. From my experience with physics I knew that one needs to invest five times more effort and time into data collection than into analyzing the data and publishing the results. As my friend had administrative and teaching duties to attend to, I took upon myself to follow up this project.

What are the most significant turning points in history *objectively* speaking? Answers to this question can be found in compilations of most-significant-milestones lists, a favorite intellectual pastime and object of diverse academic endeavors. An example was John Brockman's book *The Greatest Inventions of the Past 2000 Years.*[1] It was a collection of often-whimsical opinions from an elitist group of intellectuals, some of whom I had met as fellow faculty in DUXX—a leadership school experiment in Monterrey, Mexico. The participants had lots of fun responding to the call and produced a list of about 75 inventions that included such "exotic" milestones as the appearance of free will, ego, and the idea of an idea.

Less witty but answering the more relevant question "What are the major events in the history of life?" are lists that can be found in the National Geographic magazine, or compiled during scientific gatherings such as symposia convened by biologists and ecologists. Still closer to the question were lists in more conventional depositories of macroscopic knowledge, such as the *Encyclopedia Britannica* and the American Museum of Natural History. They both offered compilations of "Major Events in the Universe's History." I found similar lists in *Scientific American* and in a number of books by scientists.[2] To these I added Carl Sagan's celebrated *Cosmic Calendar* that matches the entire history of the Universe onto one year, pointing out dates of major events.[3]

But I also used another technique, writing letters. I wrote to over sixty Nobel Prize laureates (in physics, chemistry, and medicine) asking them for what they considered to be the twenty-five or so most significant turning points in the evolution of the world. The response was very poor. I received a handful of answers and only one complete set of milestones, from biochemist Paul D. Boyer.

In any case, I distilled fifteen *reliable* sources of data and combined them to produce what I called the set of 28 *canonical* milestones, which indeed coalesced in recent times. Appendix E lists these milestones.

The fact that these milestones were of *top-most* importance made them of *comparable* importance. I then argued that the importance of each milestone had to be proportional to the amount of change introduced but also proportional to the distance to the next milestone (an important change carries long-term consequences). Thus I was able to build a quantitative graph for the evolution of change, which I then fitted to an S-curve that had its origin at the time of the Big Bang and was only half-way completed by our times.

THE PRIMORDIAL S-CURVE

My 28 canonical milestones represent the most significant turning points in the evolution of the Universe. The emerging overall trend displays an unambiguous crowding of milestones in recent times. The thinking behind my article was that the spacing of the most important milestones could serve to quantify the evolution of change and complexity and therefore enabling us to forecast it.

It is reasonable to assume that the greater the change associated with a given milestone, or the longer the ensuing stasis, the greater the milestone's importance will be.

Importance = (change introduced) x (duration to the next milestone)

Following each milestone there is change introduced in the system. At the next milestone there is a different amount of change introduced. Assuming that milestones are approximately of equal importance, and according to the above definition of importance we can conclude that the change ΔC_i introduced by milestone i of importance I is

$$\Delta C_i = \frac{I}{\Delta t_i}$$

where Δt_i the time period between milestone i and milestone i+1.

Change per Milestone

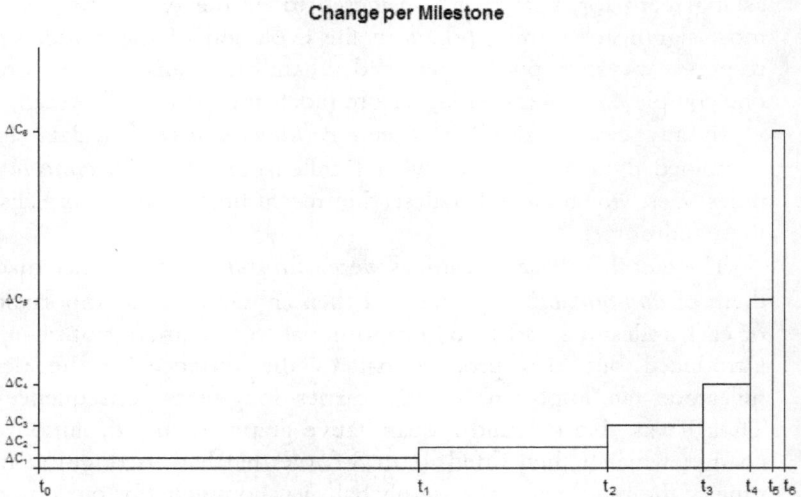

FIGURE 9.1 To the extent that milestones of equal importance appear more frequently, the respective change they introduce increases. The area of each rectangle represents importance and remains constant. The scales of both axes are linear.

My dataset has a number of weaknesses. Only twelve out of the fifteen timelines used were independent. One timeline had been given to me without dates and I introduce them myself; another consisted of my own guesses. Both were heavily biased by the other twelve in my disposal. Moreover, some data were simply weak by their origin (e.g., an assignment post on the Internet by a biology professor for his class, which is no longer accessible today.)

As a matter of fact only one timeline (Sagan's Cosmic calendar) covers the entire range (Big Bang to 20th century) with dates. A second complete set (by Nobel Laureate Paul D. Boyer) was provided to me without dates. All the other timelines coming from various disciplines covered only restricted time windows of the overall timeframe, which results in uneven weights for the importance of the milestones as each specialist focused on his or her discipline.

In any case, the grand total of all milestones came to 302 and in a histogram—Figure 9.2—revealed clusters of milestones with peaks that defined the canonical set of milestones. Present time was taken as year 2000. Within all sources of uncertainties mentioned above, I

Histogram of All Milestones

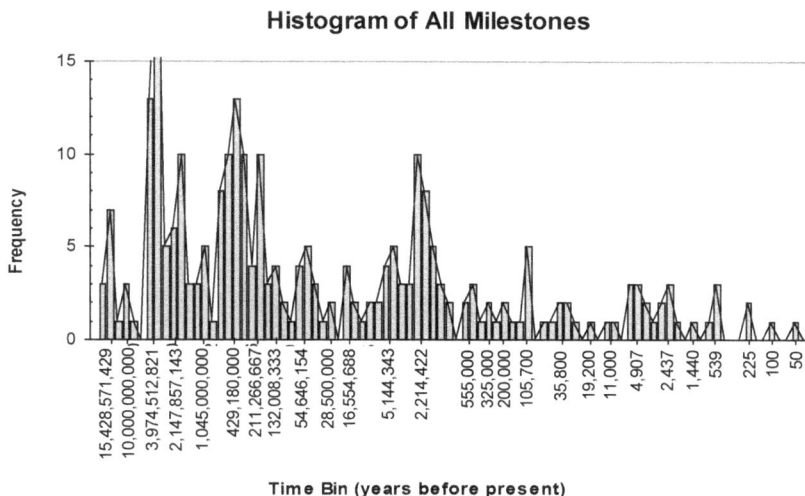

FIGURE 9.2 A histogram of all data collected with logarithmic time buckets suggested by clusters of milestone dates. The thin black line is superimposed to outline the clusters of milestones that define the dates of a canonical milestones set. On the horizontal axis we read the dates of these canonical milestones. Present is defined as year 2000. The width of each peak becomes a measure of the uncertainty on the date of the canonical milestone (and consequently also the uncertainty on the change it brings).

tried to quantitatively study the evolution of complexity and change in the Universe.

Figure 9.3 shows that the evolution of change with milestone number has so far followed an exponential-growth pattern, which could also be an S-curve or its life cycle (bell-shaped) as all three behave exponentially early on. Given that the data depict change *per milestone* I fit to the latter shown by the thick gray line. The implication of doing so is that the total amount of change in the Universe will be finite by the time the Universe ends.

The quality of the fit being a little better and the position of the last point both argue in favor of the logistic life cycle rather than the corresponding exponential. But these are weak arguments. More serious impact have the forecasts for the change expected by future milestones. Table 9.1 lists these forecasts for the next five milestones. The life-cycle fit expects milestones to begin appearing less frequently in the future whereas the exponential fit expects them at an accelerating pace. In particular, the life-cycle fit has next milestone

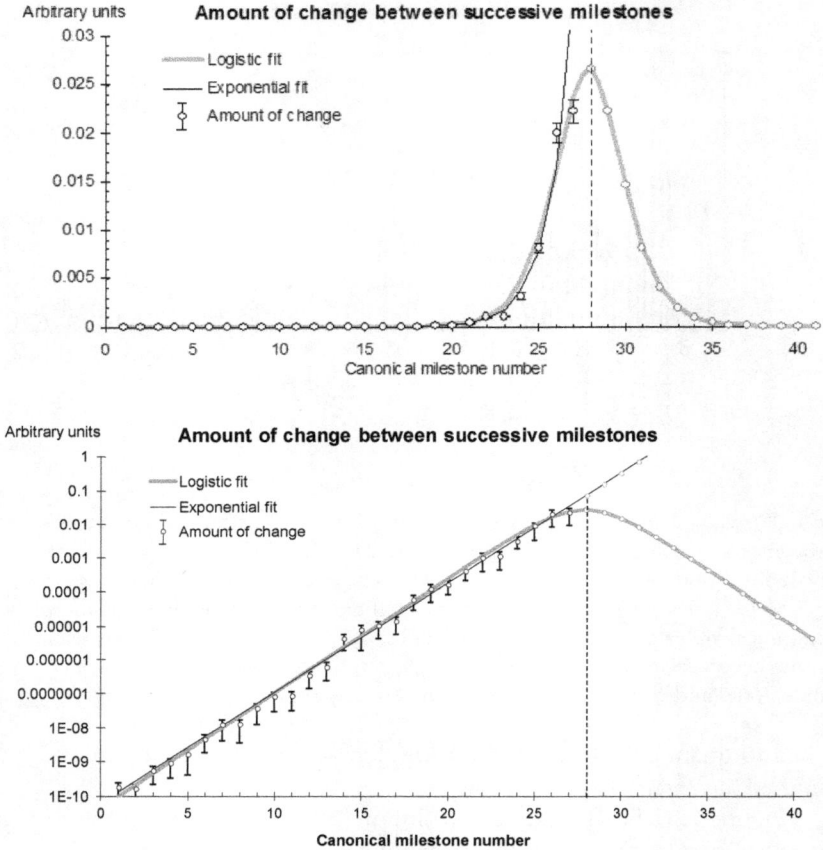

FIGURE 9.3 Exponential and life-cycle fits to the data of the canonical milestone set. The vertical axis depicting the amount of change per milestone is linear (graph at the top) and logarithm (graph at the bottom). The intermittent vertical line denotes the present. The gray circles on the forecasted trends indicate change from future milestones. The change associated with the most recent milestone, No 28, will not be known before the appearance of milestone No 29. The error bars have been calculated from the spread of entries clustered around each canonical milestone date.

appearing in 2033 and the one after that in 2078. In contrast, the exponential fit has next milestone in 2009 (remember zero was defined as year 2000 in this study), and the one after that in year 2015. By year 2022 the exponential fit forecasts a new milestone every 6 days and less than a year later infinite change will have taken

Table 9.1: Forecasts for Change as a Function of Time

Milestone number	S-curve fit		Exponential fit	
	Change*	Year	Change*	Year
29	0.0223	2033	0.1540	2009
30	0.0146	2078	0.3247	2015
31	0.0081	2146	0.6846	2018
32	0.0041	2270	1.4435	2020
33	0.0020	2515	3.0436	2021

* In the same arbitrary units as Figure 9.3

place! This spells out "Singularity"—see next section—and brings it forward by 20 years or so, but the uncertainty of this determination could easily be more than 20 years considering the crudeness of the method and the enormous timescale involved.

The life-cycle peaks in the mid 1990s. It indicates that we are presently traversing the only time in the history of the Universe in which 80 calendar years—the lifetime of people born in the 1940s—can witness change in complexity coming from as many as three evolutionary milestones. This positions us presently at the world's prime!

Coincidentally people who will partake in this phenomenon belong to the mysterious baby boom that creates a bulge on the world population distribution.** As if by some divine artifact a larger-than-usual sample of individuals was meant to experience this exceptionally turbulent moment in the evolution of the cosmos.

The large-scale life-cycle description of Figure 9.3 indicates that the evolution of change in the Universe has been following a logistic/exponential growth pattern from the very beginning, i.e. from the big bang. This is remarkable considering the vastness of the time scale, and also the fact that change resulted from very different evolutionary processes, for example, planetary, biological, social, and technological. The fitted logistic curve has its inflection point—the time of the highest rate of change—in mid 1990. Considering the symmetry of the logistic-growth pattern, we can thus conclude that the end of the Universe is roughly another 15 billion years away. Such a conclusion is not really at odds with the latest scientific

---

** It has been often argued that the baby boom was due to soldiers coming back from the fronts of WWII. This is wrong because the phenomenon began well before the war and lasted until the early 1970s. The effect of WWII was only a small and narrow wiggly dent in the population's evolution.

thinking that places the end of our solar system some 5 billion years from now.

We have obviously been concerned with an anthropic Universe here because we have to a large extend overlooked how change has been recently evolving in other parts of the Universe. Still, I believe that such an analysis carries more weight than just the elegance and simplicity of its formulation. The celebrated physicist John Wheeler has argued that the very validity of the laws of physics depends on the existence of consciousness.* In a way, the human point of view is all that counts! It reminds me of a whimsical writing I once saw on a tee shirt: "One thing is certain, Man invented God; the rest is debatable."

THE SINGULARITY MYTH

In 2005 I was asked together with four other members of the editorial board of *Technological Forecasting & Social Change* to review Ray Kurzweil's book *The Singularity Is Near*.[4] The book had received much attention with its provocative thesis that the accelerating rate of technological change would by 2045 result to the emergence of a new species—a hybrid between machines and humans—that would take over and reduce ordinary humans to what monkeys are to them. The task of reviewing the book dragged me back into the subject of accelerating change that six years earlier I had thought I was among the first to have observed only to find out later that a whole cult with increasing number of followers was growing around it. I had taken my distance from them because they sounded nonscientific and published on my own adhering to a strictly scientific approach.[5] But to my surprise the respected BBC television show HORIZON that became interested in making a program around this subject found even my publications "too speculative". In any case, for the BBC scientists the word singularity was reserved for mathematical functions and phenomena such as the big bang.

Kurzweil's book constitutes a most exhaustive compilation of "singularitarian" arguments and one of the most serious publications

---

* John Wheeler was a renowned American theoretical physicist. One of the later collaborators of Albert Einstein, he tried to achieve Einstein's vision of a unified field theory.

on the subject. And yet to me it still sounds nonscientific. Granted, the names of many renowned scientists appear prominently throughout the book, but they are generally quoted on some fundamental truth other than the direct endorsement of the so-called singularity. For example, Douglas Hofstadter is quoted to have mused that "it could be simply an accident of fate that our brains are too weak to understand themselves." Not exactly what Kurzweil says. Even what seems to give direct support to Kurzweil's thesis, the following quote by the celebrated information theorist John von Neumann "the ever accelerating process of technology…gives the appearance of approaching some essential singularity" is significantly different from saying "the singularity is near". Neumann's comment strongly hints at an illusion whereas Kurzweil's presents a far-fetched forecast as a fact.

What I want to say is that Kurzweil and the singularitarians are indulging in some sort of pseudo-science, which differs from real science in matters of methodology and rigor. They tend to overlook rigorous scientific practices such as focusing on natural laws, giving precise definitions, verifying the data meticulously, and estimating the uncertainties. Let me give here one example of what I mean by lack of scientific rigor, the reader can find more such examples in my subsequent publications on the subject.[6]

In his book *The Singularity Is Near* Kurzweil talks about the "knee" of an exponential curve, the stage at which an exponential begins to become explosive. But it is impossible to define such a knee in a rigorous way because of the subjective aspect of the word "explosive". Figure 9.4 displays four sections of the *same* exponential function. On graph (a) at the top the knee could well be at time = 70 but as we look closer it progressively moves down to time = 7 in graph (d) at the bottom. It is still the same exponential function with the vertical scale expanded.

There is no way to single out a particular region on an exponential curve because the pattern has no intricate structure. It is basically a one-parameter mathematical function that varies continually and identically from -∞ to +∞. It always grows at the same percentage rate. In contrast, the S-curve has a ceiling and a center point, which can be used as reference points.

Kurzweil's knee depends on the judgment of the observer, namely that the curve has attained an *apparently* high value. The knee can be defined as a threshold, an absolute level characterized as *high* by the majority of observers. But this is clearly a source of bias.

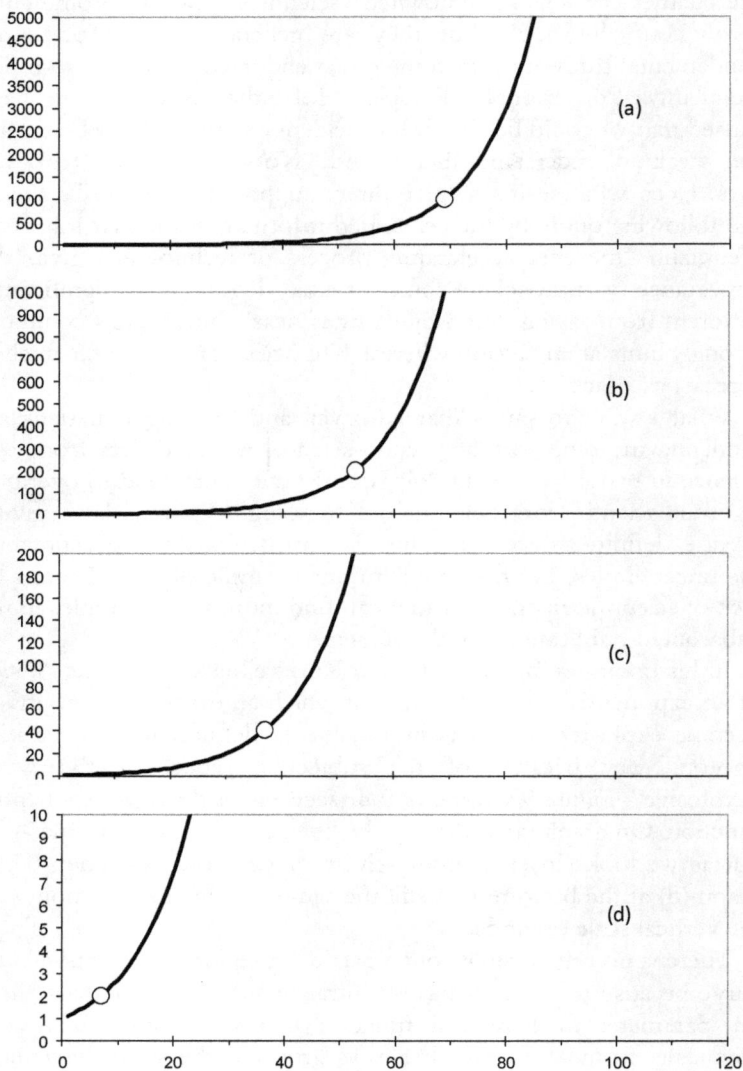

FIGURE 9.4. The same exponential is displayed with different vertical scales. Kurzweil's knee, for example the white circle, can be positioned anywhere depending on where the observer looks.

146

So I decided to present a number of science-based arguments against the possibility that the kind of singularity described by Kurzweil in his book will take place anytime during the 21st century. I concentrated on the relatively near future because horizons of several hundred years permit and are more appropriate for fantasy scenarios that tend to satisfy the writer's urge for sci-fi prose.

There Are No Exponentials. There Are only S-Curves

What supporters of the singularity notion are doing wrong is that they rely on mathematical functions for their extrapolations rather than on natural laws. The S-curve is also a mathematical function but it describes the law of natural growth in competition completely. The exponential function describes only part of a natural law. Nothing in nature follows a pure exponential. All natural growth follows the S-curve, which indeed can be approximated by an exponential in its early stages. Explosions may seem exponential but even they, at a closer look, display well-defined phases of beginning, maturity, and end, the integral of which yields an S-curve. Explosions can be described from beginning to end far more accurately by an S-curve—albeit a sharply rising one—than by an exponential. Appendix D can help us reliably estimate when exponential trends turn into S-curves.

Most readers will agree that a 25% deviation between exponential and S-curve patterns is significant because it makes it clear that the two processes can no longer be confused. This happens when the S-curve that corresponds to the exponential has reached about 20% of its ceiling level. In other words, the future ceiling that caps this growth process is about 5 times today's level.

The moment when an exponential pattern begins deviating significantly from an S-curve also defines an upper limit for Kurzweil's so called "exponential knee". If we interpret the knee as the moment when a growth process still following an exponential pattern begins having *very serious impact* on society—a subjective definition carrying large uncertainties—then there will be at least a 7-fold increase remaining before the process stops growing.

It has been theoretically demonstrated that fluctuations of a chaotic nature begin making their appearance as we approach the ceiling of an S-curve. This is evidence of the intimate relationship that exists between growth and chaos (the logistic equation in discrete form becomes the chaos equation).[7] In an intuitive way these fluctuations can be seen as a random search of the system for equilibrium around a homeostatic level.

But it has been argued that fluctuations of a chaotic nature may also precede the steep rising phase of the logistic.[8] The intuitive understanding of these fluctuations is "infant" mortality. In all species survival is uncertain during the early phases of the life cycle.

Infant mortality and common sense can help us establish a lower limit for Kurzweil's knee. Any natural growth process that has achieved less than 10% of its final growth potential cannot possibly have a serious impact on society. In fact 10% growth is usually taken as the limit of infant mortality. A tree seedling of height less than 10% of the tree's final size is vulnerable to rabbits and other herbivores or simply to be stepped on by a bigger animal.

Below we look at some real cases. They all corroborate a lower limit of the order of 10% below which the impact on society cannot be considered *very* serious.

US Oil Production

A real case is the production of oil in the United States discussed in Chapter 4 and shown again here in Figure 9.5. Serious oil production in the United States began in the second half of the 19th century. During the first one hundred years or so cumulative oil production followed an exponential pattern. But soon it became clear that the process had in fact been following an S-curve rather closely. If we try to fit an exponential function to the data, we obtain a good fit—comparable in quality with the S-curve fit—only on the range 1859-1951. As we stretch this period beyond 1951 the quality of the exponential fit progressively deteriorates. S-curve and exponential begin diverging from the 1951 onward, which corresponds to around 20% penetration level of the S-curve.*

If we were to position a "knee" *a la* Kurzweil on this exponential pattern, it could by some thinking be in the early 1930s time by which almost all horses had been replaced by cars maximizing the demand for oil. At that time the penetration level of the S-curve was around 7%. Consequently on this growth process, which seemed

---

* The fitted exponential here is not the same as the exponential that the S-curve tends to as $t \rightarrow -\infty$

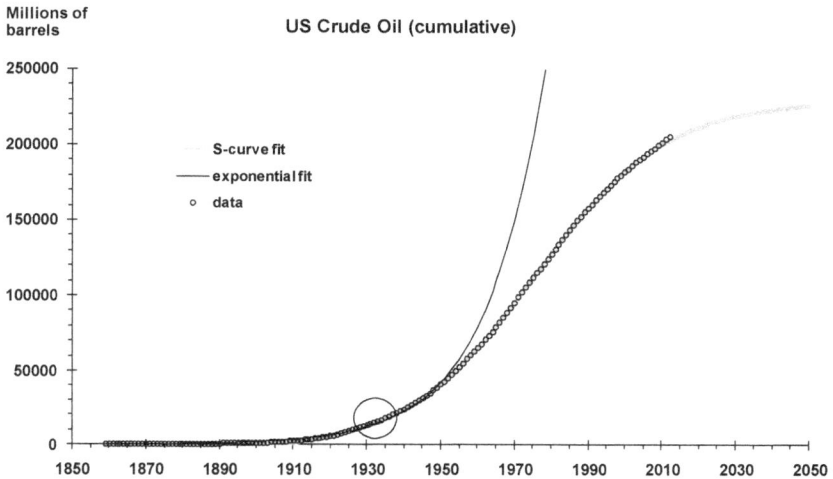

FIGURE 9.5 S-curve and exponential fit on yearly data points. The circle indicates a possible position for the "knee" of the exponential; it lies at the 7% penetration level of the S-curve.[9]

exponential for the better part of one hundred years, there is a ceiling waiting at about 14 times the knee's production level.

Moore's Law

The celebrated Moore's Law is a growth process that has been evolving along an exponential growth pattern for four decades. The number of transistors in Intel microprocessors has doubled every two years since the early 1970s. But it is now unanimously expected that this growth pattern will eventually turn into an S-curve and reach a ceiling. On page 63 of his book Kurzweil claims that Moore's law is one of the many technological exponential trends whose knee we are approaching. But he also agrees that Moore's law will reach the end of its S-curve before 2020. Moore himself agreed in 2000, "No physical thing can continue to change exponentially forever," he said and positioned an end for this phenomenon around 2015. But in 2005 he still expected three more generations (we should mention that in 1995 Moore had consistently expected 5 more generations.)

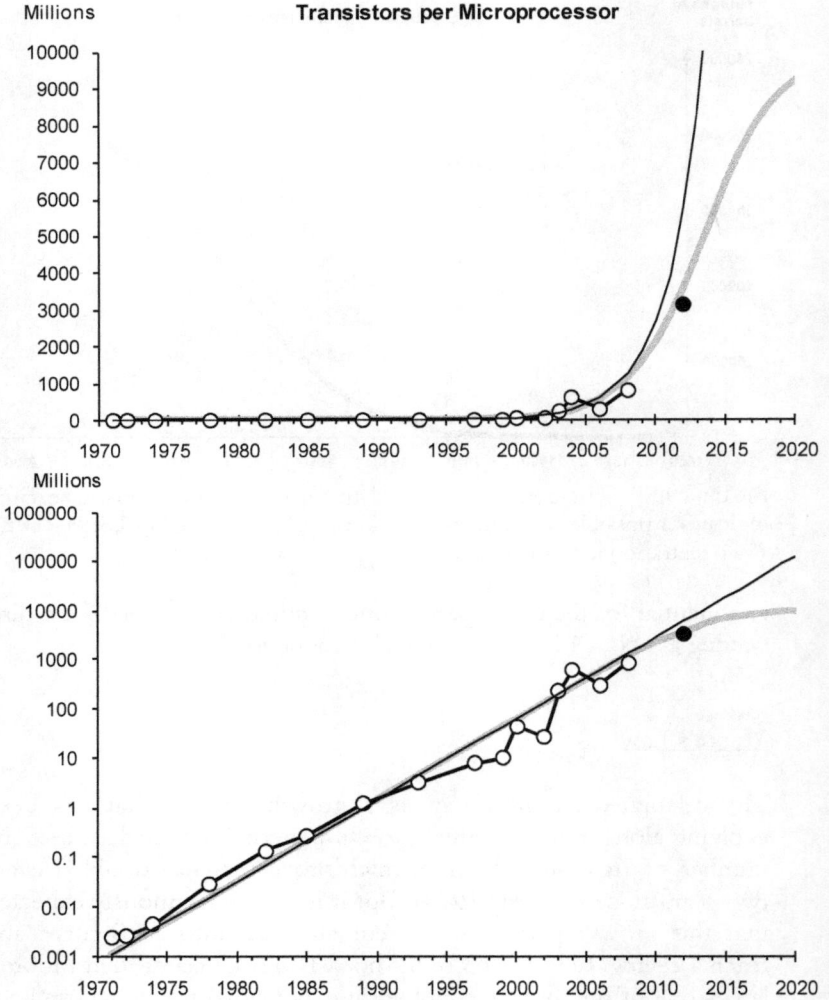

FIGURE 9.6 Moore's law with linear scale (above) and logarithmic scale (below). Exponential (black line) and S-curve (gray line) begin to diverge in the top graph around the big dotted circle (penetration of 11%) and a moment that could serve as a candidate for the "knee". The black dot indicates the Intel Poulson processor, not participating in the fit.[10]

It must be pointed out—particularly for those who claim that every time people thought a limit was reached in the past new ways were found to cram more transistors together—that in the very first formulation of Moore's law in 1965 the doubling was every year. David House, an Intel executive, raised it to 18 months, and later in 1975 Moore himself raised it to two years. These successive adjustments may not constitute proof but the fact that we are dealing with an eventual S-curve cannot be disputed. Given that we are dealing with an S-curve, the slowing down must be gradual so that three generations may bring an overall increase with respect to today's numbers by a factor smaller than $2^3 = 8$. But even if the factor is 8, today's level (which Kurzweil argues is the exponential's "knee") corresponds to around 12.5% of the S-curve's ceiling.

Figure 9.6 shows an S-curve fit on the data, which has been constrained to reach a ceiling by the late 2020s (a conservative constraint). The corresponding exponential is also shown for the sake of comparison. Intel's Poulson processor in 2012 argues in favor of the S-curve and against the exponential trend.

World Population

The evolution of the world population during the 20th century followed an S-curve in an exemplary way. Figure 9.7 shows an astonishingly good agreement between data and curve considering the variety of birth rates and death rates around the world, the multitude of stochastic processes that impact the evolution of the world population such as epidemics, catastrophes, wars (twice at world level), important climate changes, etc., and on the other hand the simplicity of the curve's description, namely only three parameters (plus a parameter for a pedestal in this case).

And yet again, the evolution of the world population has often been likened to an explosion following an exponential trend. Where could a "knee" for this exponential be positioned? Looking at Figure 9.7, any "knee" would have to be positioned before the 1980s by which time the trend significantly deviated from an exponential pattern.

By some historians the population explosion began in the West, around the middle of the 19th century. The number of people in the world had grown from about 150 million at the time of Christ to somewhere around 700 million in the middle of the 17th century. But then the rate of growth increased dramatically to reach 1.2 billion by 1850. In this case the exponential "knee" would have occurred when world population reached 8% of its final ceiling.

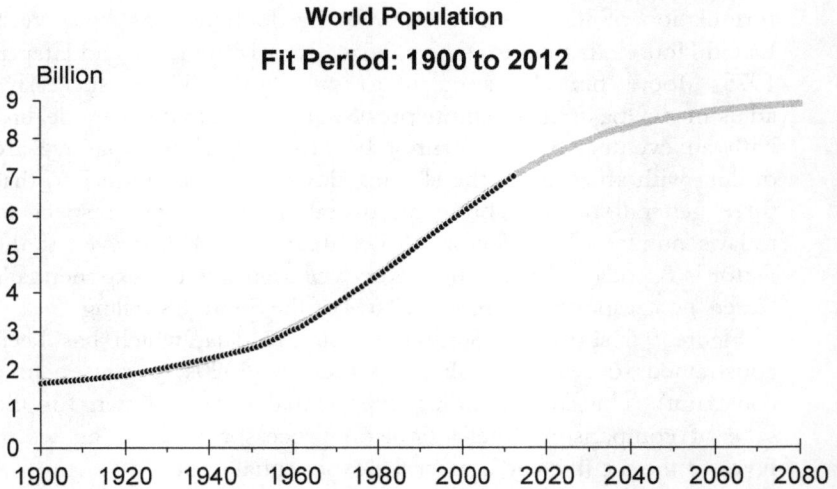

FIGURE 9.7 The evolution of the world population has followed an exemplary S-curve during at least 113 years.[11]

In the above three examples we have seen that the "knee" of the exponential curve tends to occur at a threshold situated between 7% and 13% of the ceiling of the corresponding S-curve, which translates to a factor of at most 14 between the level of the "knee" and the level of the final ceiling. This estimate is rather conservative and it corroborates a corollary of natural-growth studies, namely infant mortality.

Such growth potential on any of the variables purported to contribute to a singularity by mid 21st century would fall short of becoming alarming.

There Is already Some Evidence for Saturation

In Chapter 7 we saw that the S-curve describing the Windows family of operating systems has proceeded well beyond the center point and it is reasonable to expect an end for this growth process by the late 2020s. Appropriately this coincides with the earlier estimate for the end of Moore's law. There should certainly be expected operating systems (or the equivalent) to replace Microsoft Windows in the future, but we are facing here an upcoming lull in the rate of growth of such "explosive" processes as microchip and PC operating-system improvements. This lull should also last about 20 years, duration comparable to the duration of the rapid-growth phase of the envelope S-curve. From then (ca 2030) onward a new sequence of cascading S-curves may slowly enter the scene. But once again, we have a hard time accommodating singular events by mid 21st century due to "explosive" trends like these.

Looking at things more carefully we find that some of the very examples supporting the Singularity notion show signs of early deviation from the exponential pattern. Below are two such examples.

The US GDP

The evolution of the gross domestic product in constant dollars in America was discussed in Chapter 2 and was found to follow an S-curve. Kurzveil was too quick to pronounce its evolution as exponential. On logarithmic scale Kurzveil's straight wide band accommodated the gentle curving of the time-series data and his criterion of correlation—coefficient r^2—was close enough to 1.0. But high correlation between two curves does not mean one is necessarily a good representation of the other. Just think of two straight lines, one almost vertical the other almost horizontal; they will be 100% correlated (correlation coefficient 1.0) and yet one will be a very poor representation of the other. Figure 9.8, with logarithmic scale like that of Kurzweil, reveals that an S-curve fits better the evolution of the US GDP than an exponential one. And if we judge the fits by their Chi Square (χ^2), a criterion more appropriate than correlation coefficients, the S-curve fit comes out three times better than the exponential one (χ^2 of 1112 instead of 3250). Conclusion: the US GDP will certainly not contribute to the building up toward a singularity event around 2045 because from late 2005 onward its rate of growth has been progressively slowing down.

The U.S. GDP in Chained 2005 Dollars

GDP in billion $

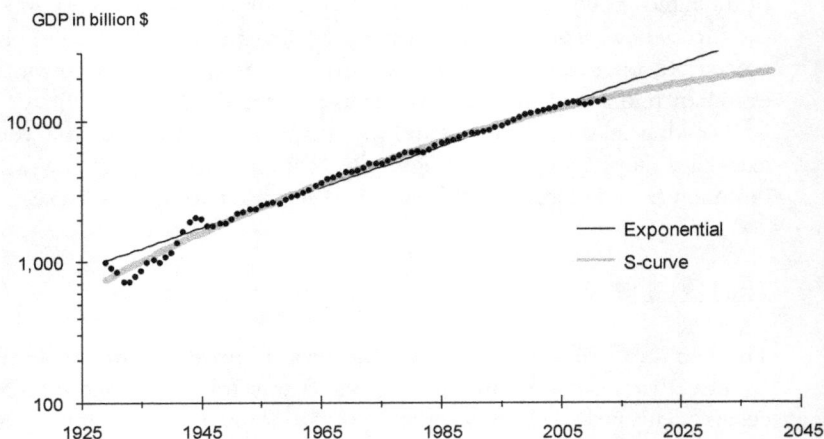

FIGURE 9.8 Exponential (thin black line) and logistic (thick gray line) fits on the evolution of the real US GDP. This graph can be directly compared with the one on Page 98 from Kurzveil's *The Singularity Is Near*.[12]

The End of the Internet Rush

One of the "explosive" variables in Kurzweil book is that of the diffusion of the Internet, but the graph on Page 79 of his book looks more like an S-curve developed half-way to its ceiling. The logarithmic version of this graph on the previous page is not quite a straight line; it has some curvature, which indicates an S-curve rather than an exponential. In fact, instead of an explosion we are witnessing the end of the Internet rush. In an article published in 2005 I demonstrate that the number of Internet users will not grow much in near future. In the US the ceiling of the S-curve has already been reached at 72% of the population, and in the E.U. it should not rise much above today's 67%. In the rest of the world the 18% of 2005 would grow to a ceiling of 33% by 2015.[13]

However, it would be unreasonable to expect the percentage of the rest of the world to remain at this low level forever. The rest of the world includes such countries as Japan, Korea, Honk Kong, and Australia where the number of Internet users is already practically at maximum. But the rest of the world also includes Africa, China, and India, where one can be certain that the number of Internet users will eventually grow by a large factor. However, it will be some time before the necessary infrastructures are put in place there to permit large-scale Internet diffusion.

For the time being one may infer that the boom we have been witnessing in Internet expansion is over. The parts of the world that were ready for it have practically filled their niches whereas the parts of the world that were not ready for it need much preparatory work (infrastructures, nourishment, education, etc.) and will therefore grow slowly.

The final percentage of Internet users may also reflect cultural differences. A percentage of 72% in the US compared to 67% in the E.U. might partially reflect missing infrastructures in some of the lesser-developed E.U. countries but most likely also reflects the different life styles. European society admits less change than American society. For example, there are fewer cars per inhabitant in Europe, and the Europeans never went to the moon. They will probably end up using the Internet somewhat less than the Americans.

We can make a rough estimate of when a follow-up Internet growth phase should be expected. To a first approximation S-curves that cascade harmoniously show periods of low and high growth of comparable duration.[14] Accordingly, and given that Internet has had a decade of rapid growth, a decade of low growth can reasonably be expected before a new S-curve begins. Contrary to the image of an explosive uncontrollable growth process we are witnessing piecemeal growth with stagnating periods in-between that offer fertile ground for control and adaptability.

Society Auto-Regulates Itself

There have been many documented cases where society has demonstrated a wisdom and control unsuspected by its members. In this section we will see four such examples portraying society as a super species capable of auto-regulating and safeguarding itself from runaway trends. This would be one more argument that a Singularity type of event is unlikely to be accompanied by disastrous consequences.

Car Accidents

We saw in Chapter 1 that the number of fatal car accidents as an annual percentage of all deaths grew exponentially during the early decades of automobile's diffusion in American society, see Figure 1.2. But this explosion did not continue indefinitely. Society enforced a tight regulation on this number limiting it at about 24 per 100,000 per year, a level that was maintained for six decades.

Another such example of society's ability to auto-regulate and safeguard itself is the spreading of AIDS in the United States.

The AIDS Niche

At the time of the writing of my first book *Predictions,* AIDS was diffusing exponentially claiming a progressively bigger share of the total number of deaths every year, and forecasts ranged from pessimistic to catastrophic. Alarmists worried about the survival of the human species. But finally the AIDS "niche" in the US turned out to be far smaller than that feared by most people.

The S-curve I fitted on the data up to and including 1988 had indicated a growth process that would be almost complete by 1994. The ceiling for the disease's growth, projected as 2.3% of all deaths was projected to be reached in the late 1990s, see Figure 9.9. In other words my conclusion at that time was that a place had been reserved for AIDS in American society just above the 2% level of all deaths.[15]

The little circles in the figure confirm the S-curve trend and the completion of this microniche by 1995. By the late 1990s questions were being raised why forecasts had overestimated the AIDS threat by so much.

There seems to have been a mechanism that limited AIDS in a natural way even in the absence of adequate medication. As if there were other, more important causes of death. This mechanism may have reflected the control exercised by American society through subconscious collective concern. The natural-growth pattern that the disease followed from its beginning correctly anticipated that AIDS would not spread uncontrollably. Eventually of course there would be effective medication for the disease and the number of victims would decline. Those who had predicted imminent doom in the absence of a miracle drug in the 1980s had failed to take into account the natural competitive mechanisms which regulate the split of the

The AIDS Niche in the US

AIDS victims as a
percentage of all deaths

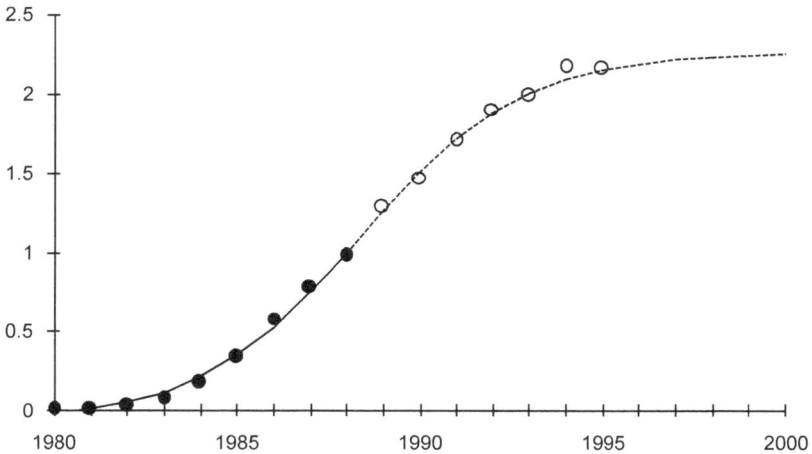

FIGURE 9.9 Deaths from AIDS in the United States. The ceiling of the S-curve fitted on data up to and including 1988 (black dots) is 2.3%. Data from years 1989–1998 (open circles) confirm that the AIDS niche in the United States was essentially completed by 1995.

overall death rate among the different causes of death, safeguarding all along an optimum survival for society.

After 1995 the number of deaths from AIDS progressively declined in what could be described as another natural process, a downward S-curve—not shown in the figure—reflecting the development of progressively effective medication. As presented Figure 9.9 demonstrates society's ability to control explosive trends and safeguard its wellbeing in the absence of effective medication and miracle drugs.

A more subtle example of society's ability to auto-regulate and safeguard itself is primary-energy substitution and the advent of nuclear energy.

Nuclear Energy

We saw in Chapter 4 that the succession of primary energies followed closely the S-curve substitution model with any deviations reflecting on social unrest (strikes) and the behavior of environmentalists.

Environmentalists have taken a vehement stand on the issue of nuclear energy. This primary energy source entered the world market in the mid 1970s when it reached more than 1% share. As we saw in Figure 4.10 the rate of growth during the first decade seems disproportionately rapid, however, compared to the entry and exit slopes of wood, coal, oil and natural gas, all of which conform closely to a more gradual rate. At the same time, the opposition to nuclear energy also seems out of proportion when compared to other environmental issues. As a consequence of the intense criticism, nuclear energy growth has slowed considerably, and it could well now follow the S-curve proposed by the model.

The worldwide diffusion of nuclear energy during the 1970s followed a rate that could be considered excessive, compared to previous energy sources. The market share of nuclear energy grew along a trajectory much steeper than the earlier *natural* ones of oil, natural gas, and coal. This abnormally rapid growth—possibly responsible for a number of major nuclear accidents—may have been what triggered the vehement reaction of environmentalists, who succeeded in slowing it down. Appropriately, as the technology matured, the number of major nuclear accidents was drastically reduced. However, environmentalists are far from having stopped nuclear energy. Ironically, under pressing concerns of CO_2 pollution, their opposition to nuclear energy had considerably weakened until the accident at Fukushima nuclear plant. I believe they will again reduce their opposition as public opinion cools off and better safety measures are put in place.

The changeable behavior of environmentalists suggests that there are other more fundamental forces at play while environmentalists behave more like puppets. These forces do not involve governments and their policies that usually become shaped after the fact in response to public outcries.

<u>World Population – The Big Picture</u>

Another example of society's wise and unsuspected ways of controlling human behavior is the slowing down in the rate of growth of the world population during the 20th century. The phenomenon has sometimes been erroneously attributed to people having become aware of the perils of Earth's overpopulation and reacted accordingly with adequate birth control. But it is only China that has imposed nationwide birth controls via legislation and that accounts for only 20% of the world's population. The main reason the world population has slowed down is that rising standards of living offer people more highly preferred things to do than having children. The flattening of the S-curve shown earlier in Figure 9.7 is very little a consequence of top-down conscious decision-making. It is mostly a consequence of subconscious bottom-up forces shaping the patterns of our lives.

But let us zoom back and consider a much greater historical window. Figure 9.10 shows world population data since the time of Christ (before that time estimates are rather unreliable). With 50-year time bins, the slowdown in the second half of the 20th century becomes hardly noticeable. A dramatic exponential pattern emerges and seems to belong to an S-curve penetrated only to 23% by year 2000 having an eventual ceiling of about 1,750,000,000,000 around year 2700. Obviously the uncertainties involved on the level of the ceiling estimated from data that cover only the beginning of the S-curve are very large. From a detailed Monte-Carlo study on error estimation we obtain up to ±75% uncertainty on this number with confidence level of 90%.[16]

The year 2700 is a far-fetched horizon date for making forecasts and such statements are more appropriate to fiction than to scientific writing, but can there be a grain of truth? At first glance such a conclusion may seem absurd by today's standards and in view of Figure 9.7. But is it really absurd? Could the S-curve of Figure 9.7 be followed up by other S-curves in the paradigm of fractals discussed in Chapter 7? Marchetti has published an article titled "On 10^{12}: A Check on Earth Carrying Capacity for Man".[17] In this article he provided calculations demonstrating that it is possible to sustain one trillion people on the earth without exhausting any basic resource, including the environment! His calculations take into account availability of resources, energy, housing, and the environment. If he is right that there is no fundamental law violated by reaching such a

World Population

Billion

Fit Period: 0 to 2000

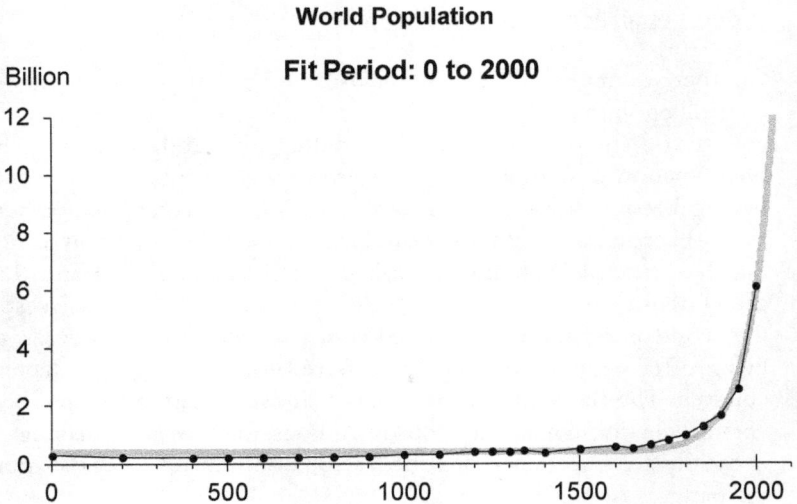

FIGURE 9.10 The world population since the time of Christ has followed the early part of an S-curve, compatible with an exponential. The slowdown during the 20th century is not visible with 50-year time bins.[18]

number, you can be sure that this will *eventually* happen (no niche in nature that could be filled to completion was ever left occupied only partially). But as we saw earlier, people's subconscious behavior during the 20th century assures us that such a thing would take place slowly, controllably, and avoiding catastrophe.

More than Just Cerebral Intelligence

This section addresses the claim by some supporters of the Singularity notion that intelligent machines will eventually take over as a new species, *posthumans*, reducing humans to the equivalent of monkeys for us today.

Intelligence according to the singularitarians is measured by the speed of calculation. I believe that the astronomical numbers of FLOPS (floating-point operations per second) forecasted by Kurzweil as the ultimate computing power, namely 10^{50} and beyond, will fall short by a large factor, at least by 25 orders of magnitude, mostly because such computing power will no longer be desired. You can get too much of a good thing, for example there is no longer demand for cars to go faster, or for pocket calculators to become thinner/smaller, something that would have never crossed the mind of early-car and early-calculator users.

But assuming unfathomable computing power becomes available, this would only be competition to our cerebral intelligence. Humans also possess physical intelligence responsible for our reproduction, growth, self-healing, and disease-fighting capabilities. Human understanding of the body's intelligence is to say the least inadequate. For example, the enteric part of our autonomic nervous system has more memory capability than the spinal cord. A mouse will function brainless to an impressive extent. Such phenomena are far from having been thoroughly investigated and understood. Where will this knowledge come from?

In any case there is a catch. If we humans were to provide the posthumans with *all* the knowledge, we would certainly refrain from giving them the power to overtake us, or build in mechanisms to prevent such an eventuality. If on the other hand, posthumans were to obtain themselves the missing knowledge, then they would need to do the studying themselves. But before posthumans begin dissecting us and putting us under microscopes—as we do with monkeys—they will first need physical bodies themselves, which they should be able to fabricate and maintain. One cannot argue that advance robotics will produce machines that can do that because there is a catch. In order for these robots to be able to move around, gather resources, and carry out research to acquire the missing knowledge they would first need to have in their system the vey knowledge they set out to obtain through studies.

On the other hand, intellectual power all by itself would not achieve much. Besides some evidence for occasional correlation, it is well known that in general among the most intelligent people you are not likely to find: the richest, the happiest, the most normal (by definition!), the best-adjusted, the most good-natured, the most trustworthy, the most creative, the best artists, the most powerful, the most popular, or the most famous. In short, fast thinking is not the ultimate desirable quality, and thinking faster is not necessarily better in an absolute sense.

All in all, posthumans enabled to develop thanks to super-computing power would certainly not pose a threat to humans by mid 21st century. In fact, I wouldn't hesitate to extend this reassuring message to the far-fetched horizon date of 2700 that we mentioned earlier.

THE HUMAN SPECIES IS NOT THREATENED

Technological achievements are manmade which by construction are designed to serve the needs of humans and one of these needs is not the extinction of the species. Of course there have been and there will always be individuals who will employ a technology in ways other than the ones intended for (notably malevolent). The scenario of the bad scientist is all too frequent, from Frankenstein to Dr. Strangelove. In addition, there is always the risk of runaway accidents and Chernobyl and Fukushima were only small such examples.

Up to now human society has averted catastrophe by reacting always in time. Strategic Arms Limitation Talks (SALT) have resulted in agreements that made a nuclear holocaust seems unlikely now. The Montreal Protocol[*] has resulted in the ozone hole getting progressively smaller and projections anticipate the hole to disappear by mid 21st century. Pollution awareness is rapidly rising around the globe; environmentalists' opposition has been restraining the diffusion of nuclear energy and stimulating technology and safety improvements both in the production of nuclear energy and the disposal of nuclear wastes. But the possibility remains, and the argument has been made, that under some circumstance in the future it may be too late for society to react. Society's good past record provides no guarantee; it could be that it simply had good luck up to now. The forces of evil may win over the forces of good and a holocaust on global scale might not be avoidable sometime in the future.

Evil—leading to annihilation—is less natural than good. A species that naturally autodestructs will be deprived of its future and will end up with no lifetime to speak off. Therefore demonstrated ability to survive any length of time is evidence that forces of good dominate over the forces of evil. But there is another phenomenon that favors the forces of good, coherence. When people sing in a crowd—typically when a singer in a concert asks for audience participation—the emerging sound is always in tune, no matter how many tone-deaf members the audience counts. Tone-deaf people—those who cannot carry a tune without randomly changing scale—may sing loud but their singing drifts over the scales randomly

---

[*] In order to protect ozone countries around the world adopted in 1987 an agreement called the Montreal Protocol to phase out ozone-depleting chemicals such as chlorofluorocarbons (CFCs).

upward or downward contributing to a general white-noise background. As loud as this background maybe, the few people who *can* sing correctly will be singing together on the same key and reinforced like this will be heard distinctly over the incoherent crowd noise. Crowds always sing in tune! A tone-deaf crowd cannot exist by the mere fact that it is *natural* for people to sing correctly, even if only a few of them can do it. The fact that the few are together while the many veer in random directions makes the difference, and the sound of the few stands out above the large incoherent noise. Similarly, the forces of good are more coherent in purpose and direction than the forces of evil. Cops will generally be ahead of rubbers because the latter do not share the common direction and purpose to the same extent. Even in global organizations of evil like Al Qaida, the coherence of action and coordination is far behind say Interpol. Of course there will be terrorist attacks, accidents, disasters, over time, some of which may be rather serious. But the survival of the species will never be at stake because naturally coordinated efforts on a global scale will have the advantage of "singing in tune" to a degree that forces of evil cannot match.

The Bitcoin

A materialization of the optimism expressed above finds itself in the foundations of the Bitcoin, which emerged in the early 2010s as an electronic cash system. Bitcoin is a purely peer-to-peer version of electronic cash that allows online payments to be sent directly from one party to another without going through a financial institution. Unlike other online (and offline) currencies, it is neither created nor administered by a single authority such as a central bank. Moreover it does not rely on trust! Of course, like any other online transaction system Bitcoin's challenge was how to protect the system against attackers.

The problem was solved using a peer-to-peer network to timestamps each transaction into an ongoing chain of proof-of-work, forming ever-increasing records that cannot be changed without redoing the work. It quickly becomes computationally impractical for an attacker to interfere if honest nodes control a majority of CPU power.

Bitcoin's inventor Satoshi Nakamoto, a mysterious hacker (or a group of hackers), has written:[19]

The network is robust in its unstructured simplicity. Nodes work all at once with little coordination. They do not need to be identified, since messages are not routed to any particular place and only need to be delivered on a best effort basis. Nodes can leave and rejoin the network at will, accepting the proof-of-work chain as proof of what happened while they were gone. They vote with their CPU power, expressing their acceptance of valid blocks by working on extending them and rejecting invalid blocks by refusing to work on them. Any needed rules and incentives can be enforced with this consensus mechanism.

Bitcoin's price moves have triggered speculation but this is not as important as the currency's ability to make e-commerce much easier than it is today. And it is all founded on the fact that the forces of evil are fewer and less coordinated than the forces of good.

IN CLOSING

The steeply-rising pattern of canonical milestones (Figure 9.3), which provides a central argument for a Singularity, resembles—and to some extent is affected by—the steeply-rising patterns of Moore's law (Figure 9.6) and of world population (Figure9.10). All three show many orders of magnitude growth along exponential trends. But Moore's law and world population have both avoided the ominously rising trend, and have done so in a *natural* way.

The last two milestones with present defined as year 2000 were:
- 5 years ago: Internet / human genome sequenced
- 50 years ago: DNA / transistor / nuclear energy

The next two such world-shaking milestones should be expected around 2033 and 2078 according to an S-curve evolution, and around 2009 and 2015 according to an exponential evolution. The former eventuality is already emerging as more plausible than the latter one because despite a steady stream of significant recent discoveries, there are still no obviously candidates for such a milestone by the end of 2013. That was not the case with the last two milestones: the significance of the Internet became clear simultaneously with its diffusion, and the significance of nuclear energy had become clear long before it was materialized.

PLAYING THE DEVIL'S ADVOCATE

Could it be that on a large scale there may be no acceleration particular to our times? Could it be that the crowding of milestones in Figure 9.3 is simply an artifact of our perception? The other day I was told that I should have included Facebook as a milestone. "It is just as important as the Internet," she told me. Would Thomas Edison have thought so? Will people one thousand years from now—assuming we will survive—think so? Will they know what Facebook was? Will they know what the Internet was?

It is natural that we are more aware of recent events than events far in the past. It is also natural that the farther in the past we search for important events the fewer of them will stick out in society's collective memory. This by itself would suffice to explain the exponential pattern of our milestones. It could be that as importance fades with the mere distancing from the present it "gives the appearance", in John von Neumann's words, that we are "approaching some essential singularity". But this has nothing to do with year 2045, 2025, today, von Neumann's time—the 1950s—or any other time in the past or the future for that matter.

EPILOGUE

Dear Dr. Theodore Modis,

Thank you for sending us the details of your recent record attempt for 'Greatest number of S-curve fits'. We are afraid to say that we are unable to accept this as a Guinness World Record.

While we certainly do not underestimate your proposal, we do however think that this item is a little too specialized for a body of reference as general as ours. We receive many thousands of record claims every year and we think you will appreciate that we are bound to favour those which reflect the greatest interest…

• • •

That is how began the response from the records management team when I applied for an entry to the Guinness Book of Records as the man who has carried out the greatest number of S-curve fits. My argument was that for the previous 28 years I had been fitting S-curves to data points of historical time series on a multitude of subjects at an average rate of about 2 – 3 per day. This amounts to something between 20,000 and 30,000 fits, which should be combined with the 40,000 fits of the Monte-Carlo study we did with Alain Debecker to quantify the uncertainties in S-curve fits.[1]

S-curves have transcended the space of professional and academic interest for me; they have become a way of life. The same thing happened to other "tools" that I received in my training. The oldest one is the scientific method summarized as observation, prediction, verification. This sequence of actions permits the use of the stamp "scientific" on a statement, which then enjoys widespread respect, but most importantly helps the one who makes the statement become convinced of its validity.

Another "tool" is evolution through natural selection, which can also be reduced to three words: mutation, selection, diffusion. Mutations owe their existence to the law which says that when something can happen, it will happen (mathematically referred to as the ergodic theorem). Mutations serve as emergency reserves; the larger their number, the higher the chances for survival. The selection phase is governed by competition, which plays a supreme role and deserves to be called the "father of everything."[2] After selection, the diffusion of the chosen mutant proceeds along natural-growth curves, smoothly filling the niche to capacity. Irregular oscillations toward the end of a growth curve may be heralds of a new growth phase that will follow.

Sustained growth and change both come in successions of S-curves. For one reason or another, transitions always take place. Change may be inevitable, but if it follows a natural course, it can be anticipated and planned for. Timing is important. In the world of business, for example, there is a time to be conservative—the phase of steep growth when things work well and the best strategy is to change nothing. There is also the time of saturation when the growth curve starts flattening out. What is needed then is innovation and encouragement to explore new directions. Our leaders may not be able to do much about changing an established trend, but they can do a lot in preparing, adapting, and being in harmony with it. The same is true for individuals. During periods of growth or transition our attitude should be a function of where we are on the S-curve. The flat parts of the curve in the beginning and toward the end of the process call for action and entrepreneurship, but the steeply rising part in the middle calls for noninterference, letting things happen under their own momentum.

The seasons metaphor offers understanding and advice on what to do given where one is on the curve. For example, GDP growth in the West generally finds itself in a fall season. This renders austerity programs more appropriate than growth-stimulus packages there. In contrast, China and India are in spring seasons, which offers growth opportunities and calls for investments.

America is becoming of age. It may no longer be the front runner it used to be in so many areas. But intellectually it is not giving an inch; not yet anyway. It is posed to hold on to about half of all Nobel-Prize awards for many years to come.

There are some clouds over the energy scene. The replacement of carbon by natural gas has fallen behind schedule and it is dragging down both environment and technology, the latter because it deprives aviation of the hydrogen-rich fuel needed for supersonic transport. The other dark cloud is nuclear energy and the amount of hydrogen it has been failing to provide. Nuclear technology needs to be brought to the frontline again in R&D and in deployment. And it must be done urgently. As the Kondratieff cycle peaks in the 2020s, we will see maximum energy consumption worldwide and it will be bad news if it is done on the back of coal and oil.

But I am optimistic. S-curves tell me that the human species is not threatened by *superhumans* or other malevolent actions. In fact, the frenetic rate of change we have been witnessing will eventually be reined in. We happen to be at the top of the world, in the sense that we are traversing a period with the most rapid rate of change ever, be it in the past or in the future. We might as well enjoy it!

There have been many endorsements of my work on S curves, some by renowned scientists and others by successful business-persons who pointed out the usefulness of the subject, be it on forecasting, strategy setting, or competition management. But the quotes below have been singled out here because they attest to the added value S-curves can bring on a gut level, beyond statistical fits and quantitative analyses.

> "*Predictions* helped me think much more clearly about the world around me."
> Thomas Dorsey, Author of *Point & Figure Charting*

> "I feel there is much wisdom in the book about life in general in a variety of ways…it provided some profound insights to me."
> Ken Ferlic, Physicist

> "[Modis'] contributions will help me become an even better dad and husband."
> Hamilton Lewis II, Market Analyst

> "[*Predictions*] A fascinating book."
> Kosta Tsipis, Professor of physics at MIT

APPENDIX A

Other Updates for "Predictions"

In this appendix we find updates of all the graphs that appeared in *Predictions* and have not yet been mentioned in this book. The figures in the next pages show the same graphs as in *Predictions*—maintaining the original numbering—with recent data superimposed. There are only few comments in connection with the recent data. The reader can find extensive discussions on these figures in *Predictions*.

The Era of Particle Accelerators

Cumulative number of
particle accelerators

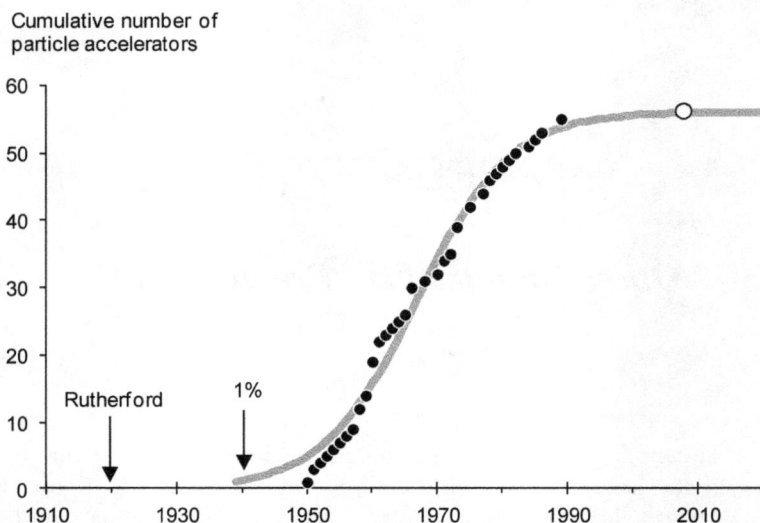

APPENDIX FIGURE 3.6 The data on particle accelerators coming into operation worldwide were fitted with an S-curve whose nominal beginning (1% level) points at War World II rather than Rutherford's famous experiment. The open circle indicates the LHC (Large Hadron Collider) that came into operation at CERN in 2008.[*]

---

[*] The data come from Mark. Q. Barton, *Catalogue of High Energy Accelerators*, Brookhaven National Laboratory, BNL 683, 1961, and J. H. B. Madsen and P. H. Standley, *Catalogue of High-Energy Accelerators* (Geneva: CERN, 1980, plus updates).

Contemporary Geniuses

Burton Richter (1931-   )

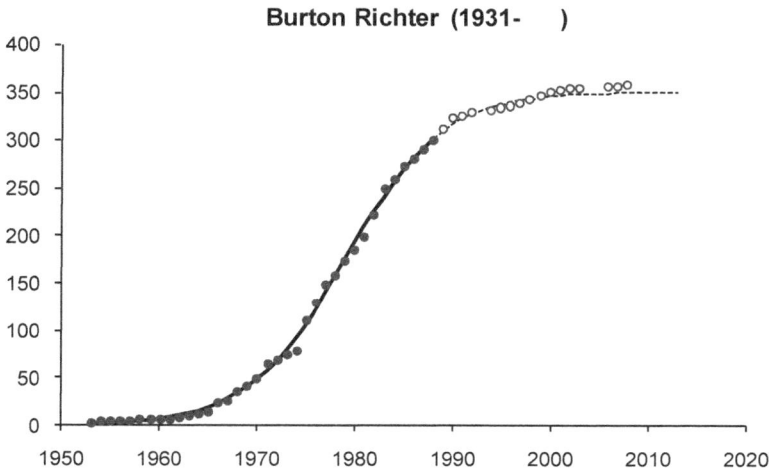

Gabriel Garcia Marquez (1928-   )

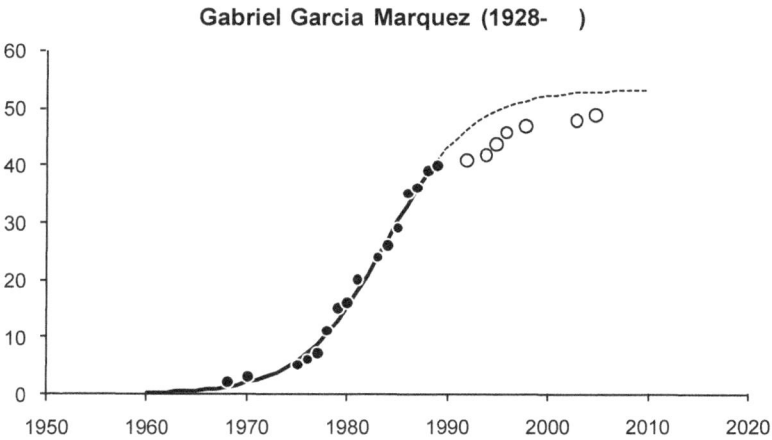

APPENDIX FIGURE 4.5 (this page and the next page). The creativity curves of four contemporary individuals. For each case we see the cumulative number of works and the corresponding S-curve as determined by a fit. The dotted lines are extrapolations of the curves. Under ordinary circumstances these people should achieve the 90% levels indicated. The small circles denote recent data.

Contemporary Geniuses (cont.)

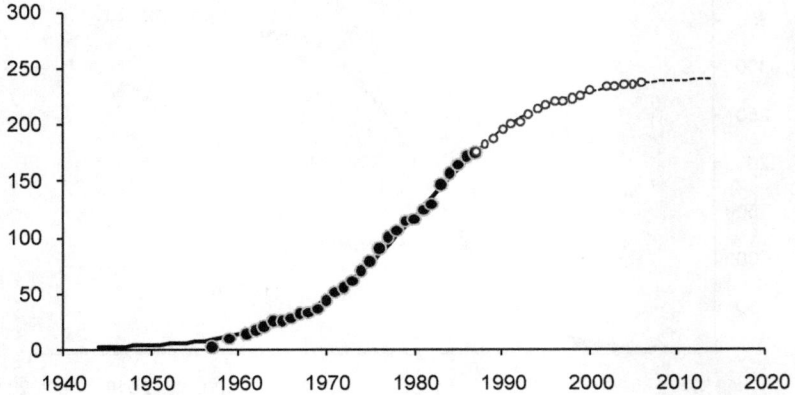

Scientific publications

CARLO RUBBIA (1934-)

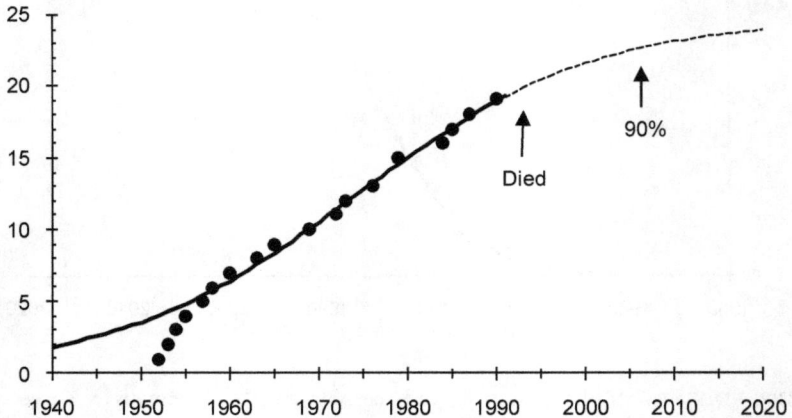

Films directed

FEDERICO FELLINI (1920 - 1993)

The S-curves as drawn in these graphs are based on data up to 1991. The little circles on the graphs tell what happened afterward. There are no surprises in the way three out of the four individuals pursued their careers. In fact, the two physicists followed their respective curves very closely, the writer followed his curve fairly closely, but the movie director did not follow it at all. Fellini died in 1993 before making any more movies.

Arguing that the more artistic the individual the more unpredictable his or her career, would be jumping to conclusions. The predictions for the two physicists turned out more accurate probably because of the amounts of data involved. There is an order of magnitude more data for them than for Garcia Marquez, and high statistics reduce uncertainties. But for Fellini the story is different. I would argue that Fellini was deprived of the chance to complete his work. It may be interesting to research into Fellini's later life for plans and scenarios of films that he did not have the time to realize. It sounds ironic— due to the age difference—but in a way Fellini's death was more premature than Mozart's.

The Death Rate Is No Longer Declining

Annual number of deaths
per 1,000 Americans

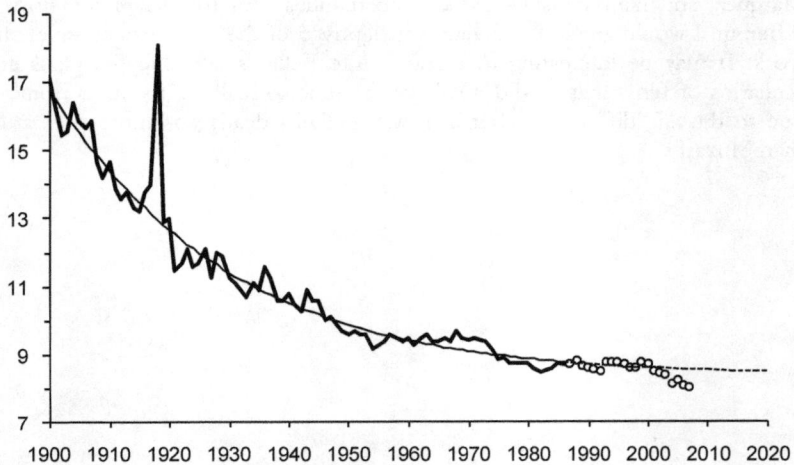

FIGURE 5.2 The death rate in the United States has practically stopped declining. The fitted curve is an exponential, or, seen differently, the final part of an upside down S-curve. World War I shows up distinctly. World War II is less visible because it lasted longer and claimed victims in smaller annual percentages of population. The small circles show what happened in the twenty years that followed the determination of the curve. The agreement is very good.*

---

* The data come from the *Statistical Abstract of the United States*, US Department of Commerce, Bureau of the Census, 1986–91; also from the *Historical Statistics of the United States, Colonial Times to 1970*, vols. 1 and 2, Bureau of the Census, Washington, DC, 1976

We Are Becoming Information Workers

Percentage of
all workers

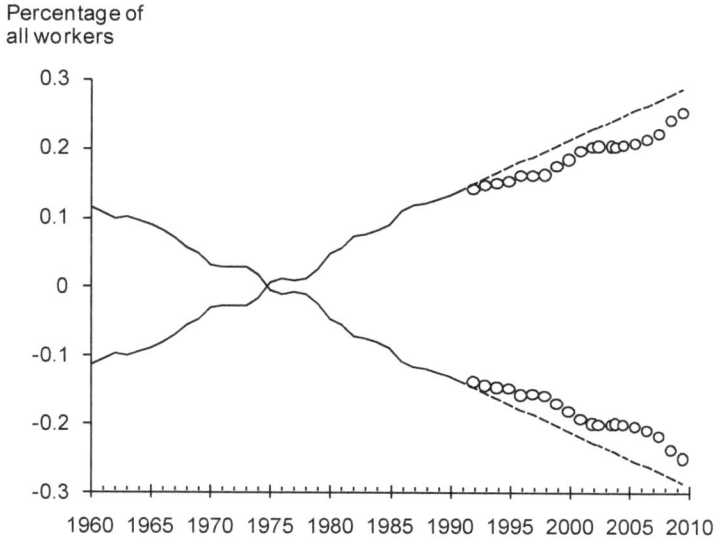

Women Executive Have not yet Reach Equality with Men

Percentage of
all executives

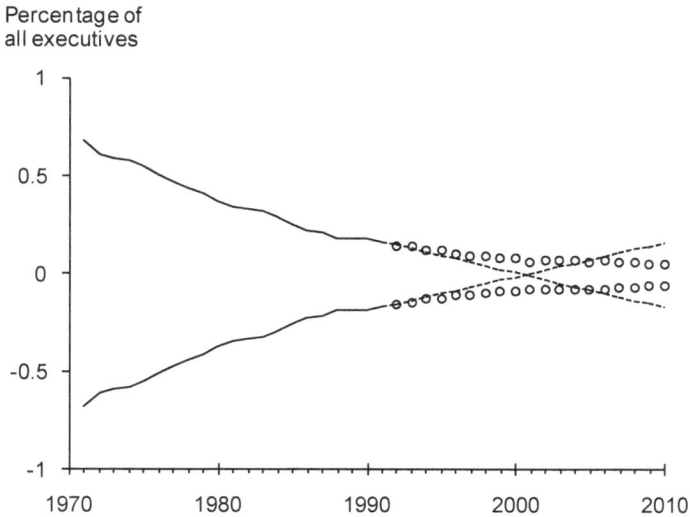

APPENDIX FIGURE 6.3 Two substitutions displaying a natural character (straight line on logistic vertical scale). At the top, the growing information content in jobs displaces manual labor. At the bottom, the percentage of women among executives projected to reach that of men by the year 2000. The little circles indicate what happened since the trends were originally determined.

The little circles tell the story of what happened during the twenty years that followed my original study. The graph at the top shows that the number of information workers kept increasing to the detriment of the number of noninformation workers as anticipated. The little circles follow closely the natural-growth pattern. But the substitution proceeded at a somewhat slower rate.*

The graph at the bottom shows a similar discrepancy. The 50-percent point for the substitution of women for men executives had been estimated to be around the year 2000. The little circles seem to indicate that this point may be reached much later. In fact it looks more reasonable now to expect that men and women may end up splitting the executive scene fifty-fifty. The women-for-men substitution process may very well be aiming at a ceiling at 50 rather than 100 percent of the executive market. After all it would be an equitable attribution of quota.*

---

* The data come from Employment and Earnings, Bureau of Labor Statistics, US Department of Labor.

Competition between Transport Infrastructures in the United States

Percentage of total length

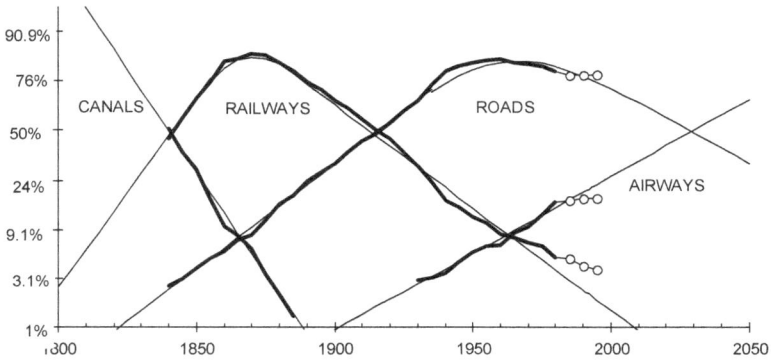

APPENDIX FIGURE 7.1 The sum total in mileage among all transport infra-structures is split here among the major types. A declining percentage does not mean that the length of the infrastructure is shrinking but rather that the total length is increasing. Between 1860 and 1900 the amount of railway track increased, but its share of the total decreased because of the dramatic rise in road mileage. The little circles show a deviation between the trajectories predicted twenty-five years ago and what happened in the following fifteen years.[*]

The evolution of airways fell short of the forecasted trajectory thus favoring of the other two infrastructures. As we saw in Chapters 3 and 4 traveling by air has been growing less rapidly than expected probably because aviation technology has not yet adopted a more hydrogen-rich fuel such as natural gas or hydrogen.

---

[*] The original graph in *Predictions* had been adapted from a graph by Nebojsa Nakicenovic in "Dynamics and Replacement of US Transport Infrastructures," in J.H. Ausubel and R. Herman, eds, *Cities and Their Vital Systems, Infrastructure Past, Present*, and Future, (Washington, DC: National Academy Press, 1988) by permission of the publisher.

The Stable Elements Were Discovered in Clusters

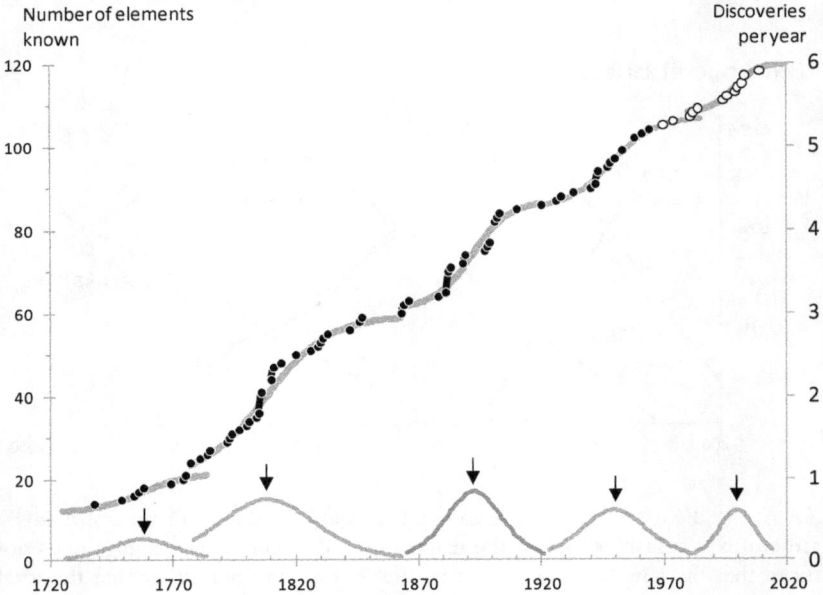

FIGURE 7.3 The black dots indicate the number of stable chemical element known at a given time. The open circles are recent data. The S-curves are fits over limited historical windows, while the gray bell-shaped curves below represent the corresponding life cycles. The arrows point at the center of each cluster hinting at some periodicity (average distance of 60.5 years).[*]

The recent data (open circles) followed the continuation of the fourth S-curve for a while but later they seem to have followed a yet another even smaller S-curve. This last cluster concerns unstable elements synthesized in laboratories with half-lives down to the milliseconds, sometimes produced in quantities of fewer than 100 atoms. What constitutes a new element in such circumstances has become an increasingly important question. The succession of cycles in element discovery must be considered approaching an end, which is corroborated by the fact that the life cycles have gotten shorter (see discussion on the fractal aspect of S-curves in Chapter 7).

It is noteworthy that the average time between clusters—60.5 years—resonates with the Kondratieff's cycle.

---

[*] The data came from The American Institute of Physics Handbook, 3rd ed. (New York: McGraw-Hill) and Wikipedia.

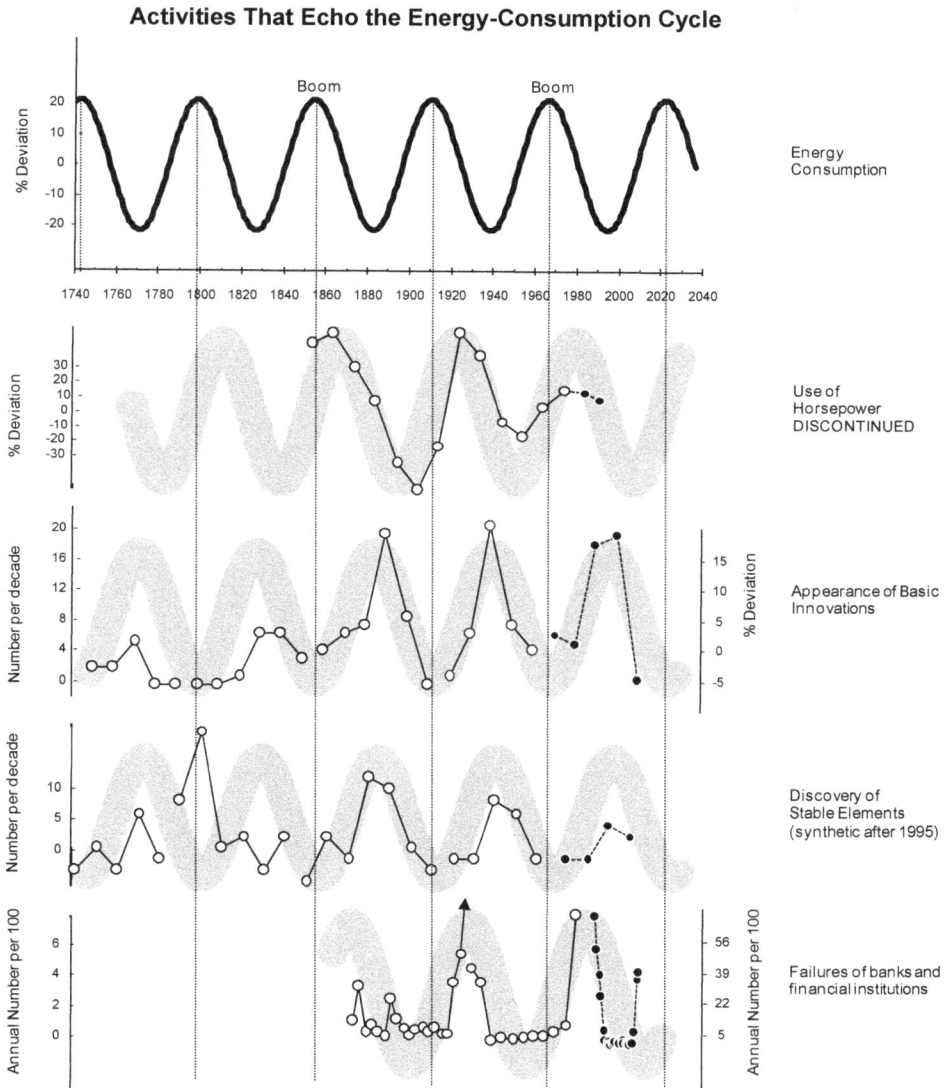

Activities That Echo the Energy-Consumption Cycle

FIGURE 8.2 On the top, an idealized energy consumption cycle is used as a clock. The data concerning the use of machines represent the percent deviation from the fitted trend. There are two sets of data for innovations: before 1960 from Gerhard Mensch's book *Stalemate in Technology*, the update points refer to the appearance of patents as a percent deviation from the growing trend of patents. There are three sets of data for bank failures: bank suspensions before 1933, banks closed due to financial difficulties (1940-1985), and percentage of banks losing money (1990-2010).

The black dots connected with intermittent lines in Figure 8.2 show what happened since the drawing was first published in 1992. Each trend has in general followed the predicted course.

Remarks on the updates:

- The time series on the use of horsepower was discontinued in 1992.
- The continuation of the appearance of basic innovations uses as a proxy the percentage deviation from the growth trend of the number of patents issued by the US Patent Office.
- The discovery of stable elements in the late 1990s loses some credibility when new discovery claims are based on fleeting observations of "elements" consisting of only 2 or 3 of atoms.
- Bank failures from 1990 onward show the percentage of banks losing money.

Activities That Echo the Energy Consumption Cycle

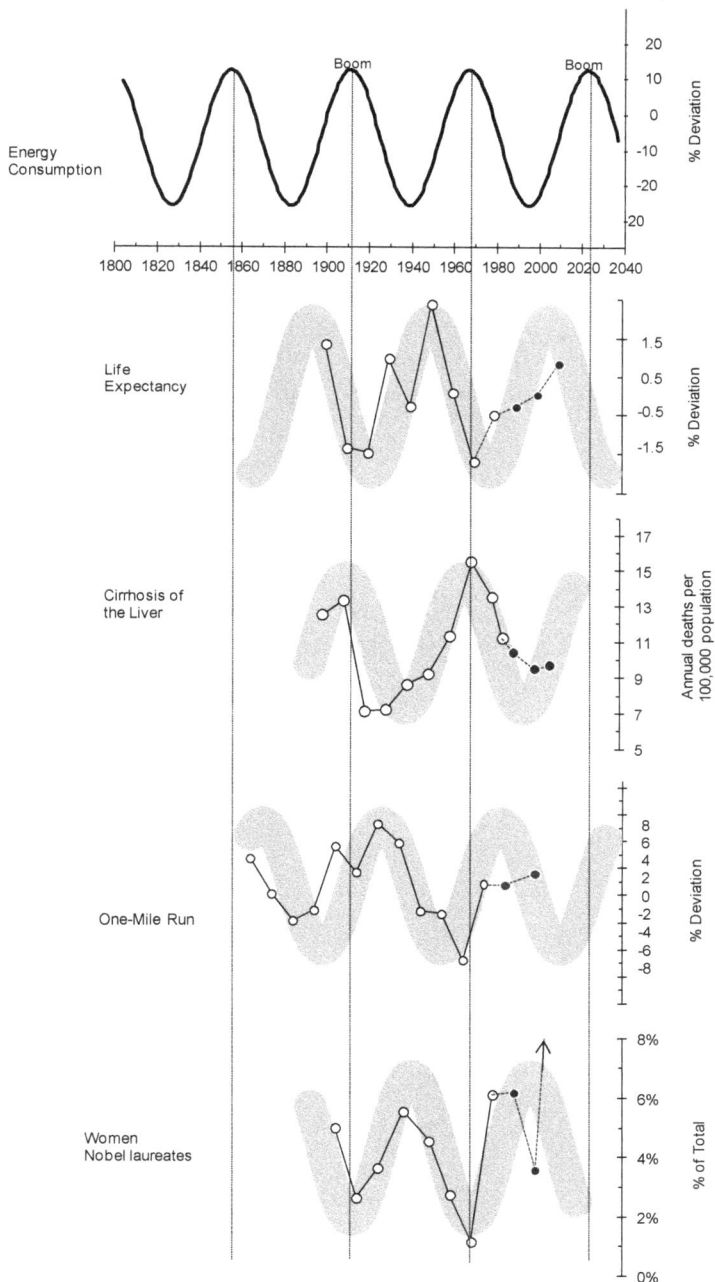

FIGURE 8.4 (previous page) The graphs on life expectancy and the one-mile-run record have been obtained in a way similar to the energy consumption, namely, as a percentage deviation of the data from a fitted trend. The number of women Nobel laureates is expressed as a percentage of all laureates in a decade.* For the case of cirrhosis the annual mortality is used.

The black dots connected with dotted lines in Figures 8.4 tell the story of what happened during the twenty years that followed. There is general agreement with the trends. As expected, life expectancy and the record breaking of the one-mile run both increased while cirrhosis-of-the-liver victims declined. Feminism—as expressed by the women content among Nobel Laureates—is also in agreement with the predictions. The most recent data point (12.5%) is off the scale, but the average of the last two data points is 8% in fair agreement with the gray trend.

---

* The graph for women Nobel Laureates has been adapted from Theodore Modis, "Competition and Forecasts for Nobel Prize Awards," *Technological Forecasting and Social Change*, vol. 34 (1988): 95-102. More recent data come from the Nobel Foundation.

Morbid Activities Also Resonate with the Same Rhythm

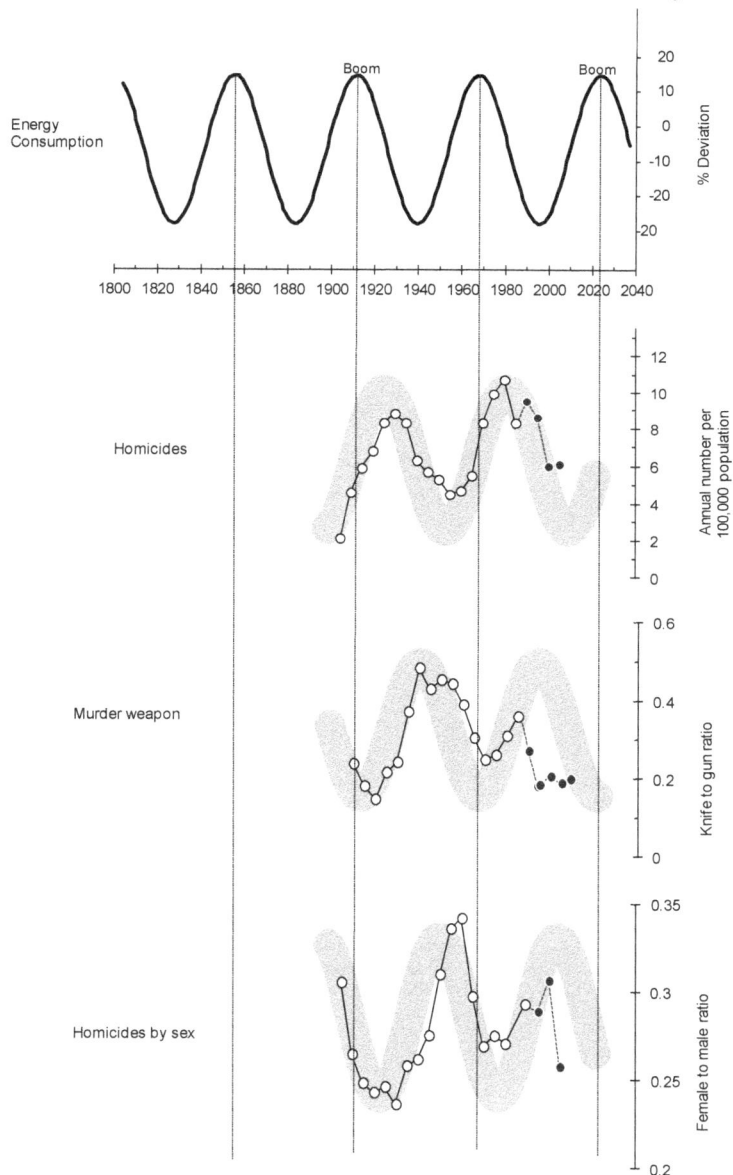

FIGURE 8.5 Below the energy consumption clock we see the annual rate of homicides. In the middle, the ratio of knife to gun as murder weapon, and on the bottom the ratio of women to men as the victim. The black dots connected by an intermittent line represent updates.

The only trend that deviated significantly from the forecasted course (i.e. declined too early) is the choice of the murder weapon (second graph). The ratio knives to guns dropped by a factor of two as guns gained in popularity, which was forecasted to happen during boom years not there yet (next boom is due in the 2020s). One may speculate that the precautious economic recovery in America during the period 1991-1998 may have "confused" murderers into using the gun prematurely, instead of the knife that would have been more "natural" at that time.

A Periodic Oscillation Recorded over Five Centuries

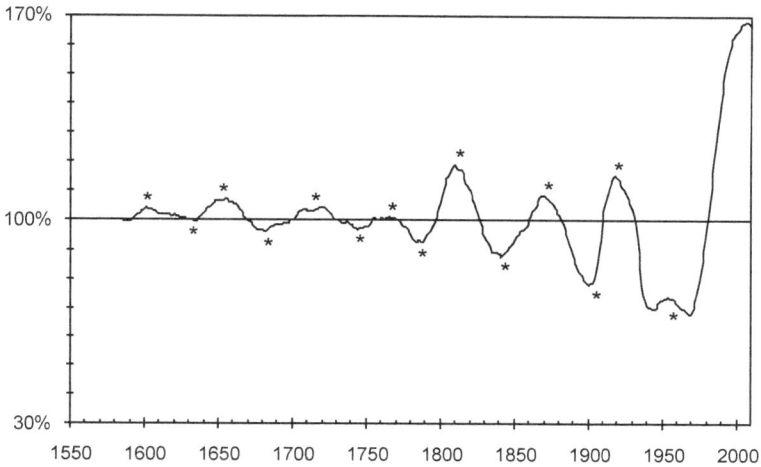

FIGURE 9.4 The U.K. Wholesale Price Index smoothed over a rolling 25-yearperiod with respect to a 50-year moving average. This procedure washes out small fluctuations and reveals a wave. The periodicity turns out to be 55.5 years. The little stars point out peaks and valleys.[*]

The cyclical pattern during the last twenty years continued as had been forecasted but an update of the original graph on a point-by-point basis is not appropriate here because of the averaging techniques involved.

---

[*] Such a graph was first published by Nebojsa Nakicenovic, "Dynamics of Change and Long Waves," report WP-88-074, June 1988, International Institute of Applied Systems Analysis, Laxenburg, Austria.

Plywood Filled a Niche in the Construction Industry

Billions of
square feet

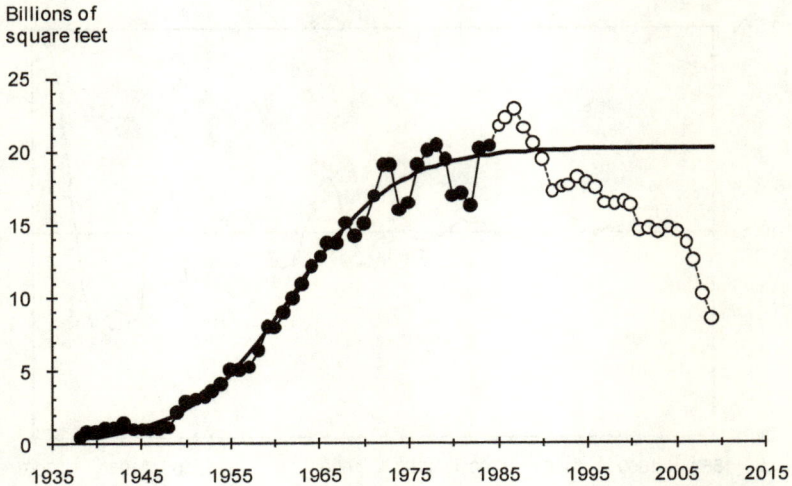

APPENDIX FIGURE 10.1 Annual plywood sales in the United States. Significant fluctuations around the smooth S-curve pattern appear when the ceiling is approached. The small circles show what happened in the twenty years that followed.[*]

The decline in plywood use from the late 1980s onward is due to the replacement of plywood by new synthetic materials on the market with competitive advantages.

[*] Adapted from a figure by Henry Montrey and James Utterback in "Current Status and Future of Structural Panels in the Wood Products Industry," *Technological Forecasting and Social Change*, vol. 38 (1990): 15-35. Recent data points come from the *Statistical Abstract of the United States*, US Department of Commerce, Bureau of the Census.

Temporary "Chaos" in US Coal Production

Millions of tons

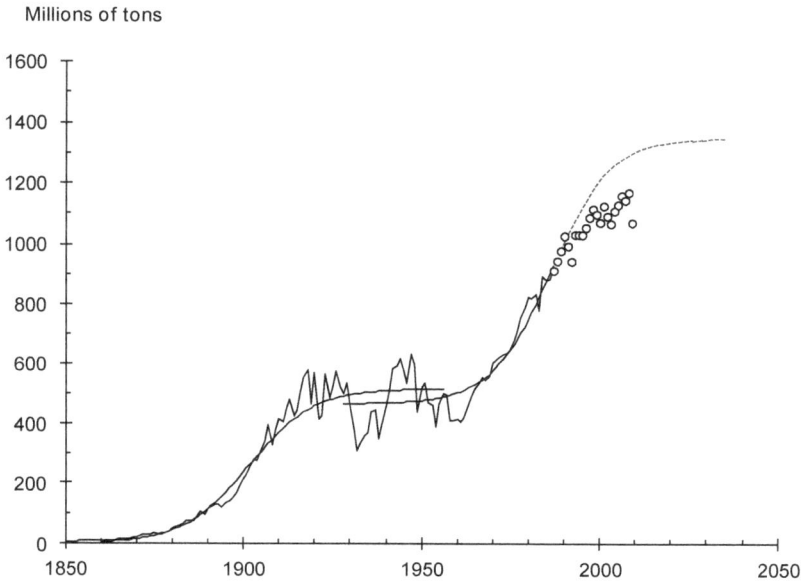

FIGURE 10.3 Annual production of bituminous coal in the United States. The two S-curves are fits to the data of the respective historical periods. The interim period shows large fluctuations of chaotic nature. The small circles show what happened during the twenty years following the original prediction.[*]

Recent data (open circles) closely followed the projected S-curve for five years and then began deviating from the projection. All in all, however, considering that we are dealing with a 20-year future horizon, the forecast can be considered acceptable.

[*] The data come from the *Historical Statistics of the United States, Colonial Times to 1970*, vols. 1 and 2. Bureau of the Census, Washington DC, and the *Statistical Abstract of the United States*, US Department of Commerce, Bureau of the Census.

Ineffective Government Decisions

Percentage of all
available oil

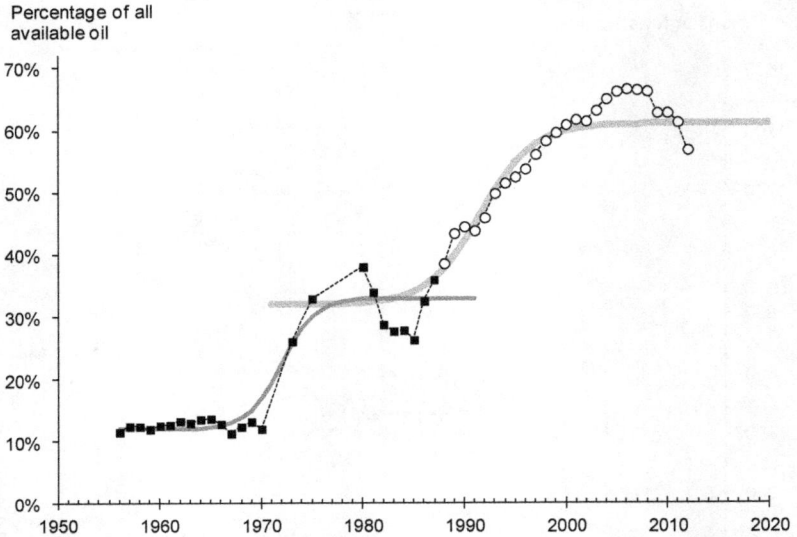

FIGURE 11.2 United States oil imports as a percentage of all available oil—imports plus domestic production. A quota on imports had to be relaxed in 1970, and the ambitious energy-independence project launched at the same time proved ineffective. The S-curves are not fits to the data but simply idealized natural-growth scenarios. The small circles represent recent data and indicate the opening up of a follow-up niche.*

---

* The data come from the *Historical Statistics of the United States, Colonial Times to 1970*, and from the *Statistical Abstract of the United States*, US Department of Commerce, Bureau of the Census, Washington DC.

APPENDIX B

Mathematical Formulation of S-curves and the Procedure for Fitting Them on Data

The behavior of populations growing under Darwinian competition has been the object of much discussion in the scientific literature.[1-6] Below are simplified formulation for the cases considered in this book

The Malthusian Case: One Species Only

An illustrative example of this case is a population of bacteria growing in a bowl of broth. The bacteria transform the chemicals present in the broth into more bacteria. The rate of this transformation is proportional to the number of bacteria present and the amount of transformable chemicals still available.

All transformable chemicals will eventually become bacteria. One can therefore measure broth chemicals in terms of bacterial volume. If we call $N(t)$ the number of bacteria at time t, and M the amount of transformable chemicals at time 0 (before multiplication starts), the Verhulst equation can be written as

$$\frac{dN}{dt} = a_x N \frac{(M - X)}{M} \tag{1}$$

and its solution is

$$N(t) = \frac{M}{1 + e^{-\alpha(t-t_o)}} \tag{2}$$

with b a constant locating the process in time.[4]

We can manipulate mathematically Equation (2) in order to put it in the form

$$\frac{N(t)}{1 - N(t)} = e^{a(t-t_o)} \tag{3}$$

Taking the logarithm of both sides, we obtain a relationship linear with time, thus transforming the S-shaped curve of Equation (2) into the straight line of the type: $at + b$. The numerator of the left side of Equation (3) is the new population while the denominator indicates the space still remaining empty in the niche. In the case of one-to-one substitutions, the numerator is the size of the *new* while the denominator is the size of the *old* at any given time. If on the vertical logarithmic scale we indicate the fractional market share instead of the ratio *new / old*, we obtain the so-called logistic scale in which 100 percent goes to plus infinity and 0 percent to minus infinity.

M is often referred to as the niche capacity, the ceiling of the population $N(t)$ at the end of growth. In order to solve Equation (1), M must be constant throughout the growth process, but this requirement can be relaxed for market shares in one-to-one substitutions because M is equal to 100 percent.

The Volterra-Lotka Equations: Two Competitors in a Niche

Even though the case of many species in the same niche has been studied in detail, for practical reasons in industrial applications we generally consider only two competitors. All many-competitor situations can be reduced to a two-competitor picture by considering the competitor of interest and grouping all the others together. This formalism is ideal when there are indeed two main competitors in the market, as with the example we saw in Chapter 8 of early mobile phones in Greece. It is also a good approximation when the market leader happens to lead all others by a large margin. The case of many small players may require further segmentation of the market into microniches.

Equation (1) can be re-written for two competitors X and Y as follows:

$$\frac{dX}{dt} = a_x X - b_x X^2 \tag{4}$$

$$\frac{dY}{dt} = a_y Y - b_y Y^2 \tag{5}$$

An interaction between the two competitors can be expressed in general terms via two coupling constants c_{xy} and c_{yx}, which are used to transform Equations 4 and 5 into a system of coupled differential equations:

$$\frac{dX}{dt} = a_x X - b_x X^2 + c_{xy} XY$$

$$\frac{dY}{dt} = a_y Y - b_y Y^2 + c_{yx} XY$$

This system of equations is usually referred to as the Volterra-Lotka system of equations and contains all fundamental parameters that determine the rate of growth—namely, the power of each "species" to multiply (a is related to attractiveness), the limitation of the niche capacity (b is related to the niche size), and the interaction with the other "species" (c is the coupling). The signs of c_{xy} and c_{yx} determine the type of competition according to the definitions of Chapter 8.

This system of equations is non-linear and cannot be solved analytically. However, solutions can be found numerically. The interested reader can find complete technical details for this procedure in *Natural Laws in the Service of the Decision Maker*.[7]

Logistic vs. Chaos Equations

The natural-growth equation (otherwise known as the Verhulst or the logistic equation) is intimately related to the equation that gives rise to states of chaos.[8] Equation (1), which gives rise to the S-curve, can be written as:

$$\frac{dX}{dt} = aX(M - X)$$

where a and M constants .

The *chaos equation*, otherwise known as the *logistic discrete equation*, is:

$$X_{n+1} = rX_n(1 - X_n)$$ where r is a constant.

The chaos equation is essentially the same as the logistic equation but in a discrete form. The two equations may be strikingly similar but the former is solved via integration and its solution gives rise to the smooth S-shaped logistic pattern, whereas the latter is solved via iteration and its solution gives rise to states of chaos for $r > 3.7$. The former emphasizes the presence of a trend and has become the tool to describe natural growth. The latter emphasizes the lack of trend and has become the tool to describe chaos. The chaotic fluctuations appear on what corresponds to the ceiling of the logistic after the upward trend has died down, as shown in the text in Figure 5.1 and repeated below for the sake of convenience.

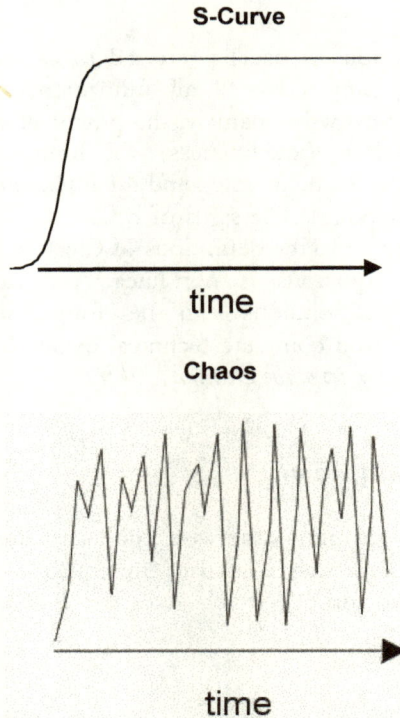

S-Curve

time

Chaos

time

APPENDIX B FIGURE 1   At the top we see the solution of the natural-growth equation giving an S-curve. At the bottom is the corresponding solution of the chaos equation giving mostly erratic fluctuations.

It has also been shown that if the logistic-growth equation is first solved and then made discrete, chaotic-type fluctuations can be expected before as well as after the curve's steep rise.[9] The pattern thus obtained consists of a cascade of growth steps giving rise to an alternation of orderly jumps and chaotic plateaus reminiscent of the per capita energy-consumption picture we saw in Figure 3.13.

Fitting and S-curve on a Set of Data Points

There are many ways to fit a mathematical function onto a sequence of data points. The method used extensively in this book, and the one likely to produce the best results, involves iterations using a computer program which tries to minimize the following sum

$$\sum_i \frac{(F_i - D_i)^2}{W_i}$$

where F_i is the value of the function, D_i is the data value and W_i is the weight we may want to assign, all at time t_i. The function is originally evaluated unintelligently by assigning arbitrary starting values to the parameters of F, but the program performs a trial-and-error search through many iterations to determine those values for which the sum becomes as small as possible.

For S-curve fitting, the function F takes the form of Equation (2), while for one-to-one substitutions it takes the form (3) or simply the straight-line expression: $at + b$.

By assigning weights to the different points we can force the computer to "pay attention" to some data points more than others, if for whatever reason there are doubts about the validity of the data.

The program we used proved quite robust in the sense that changing weights and starting values affected only the computing time but not the final results. Still, we found variations and correlations between the parameters a, b and M, particularly in the instance of the Malthusian case. In Appendix C we present some results of an extensive study on the error sizes to be expected and the correlations between the three parameters of an S-curve determined from a fit.

The interested reader can find complete technical details for the above fitting procedure in *Natural Laws in the Service of the Decision Maker.*[7]

Fitting an S-curve with a Variable Ceiling

It is often encountered that the niche into which species are growing does not remain constant over time but increases, if gently. This typically happens when the niche in question consists of a market, a population, or a living organism that itself grows with time. In such cases M is not a constant in Equation 1, which now can no longer be solved analytically but must be approximated with iterative numerical techniques. M may be increasing with time or with X, to a first approximation linearly as follows:

$$M = at + b \qquad \text{or}$$
$$M = aX + b$$

we will choose here the second case because it results in a finite growth of M over time (in fact it can be shown that it is S-shaped like X).

The logistic equation can then be rewritten as:

$$\frac{dX}{dt} = \alpha X \left(1 - \frac{X}{M}\right) = \alpha X \left[1 - \frac{X}{(aX + b)}\right]$$

where α is not the same constant as in Equation 1 and certainly different than a.

By the definition of a mathematical derivative, the centered discrete derivative for equally-spaced data points can be approximated as:

$$\frac{dX}{dt} \cong \frac{\Delta X}{\Delta t} = \frac{X_{n+1} - X_{n-1}}{2\Delta t}$$

which for $\Delta t = 1$ can transform logistic equation to the following:

$$X_{n+1} = 2 \left[\alpha X_n - \frac{\alpha X_n^2}{(aX_n + b)}\right] + X_{n-1}$$

This equation can be easily programmed into an Excel worksheet. Subsequently the sum of differences squared can be calculated with the values X_{n+1} subtracted from the corresponding data point. The Solver can then be invoked to minimize this sum by varying a total of four parameters, namely α, a, b, and X_o, the last being the starting values for the iteration.

The interested reader can find complete technical details for the above fitting procedure in *Natural Laws in the Service of the Decision Maker*.[7]

APPENDIX C

Expected Uncertainties on S-Curve Forecasts

Alain Debecker and I undertook an extensive computer simulation, (a Monte Carlo study) to quantify the uncertainties on the parameters determined by S-curve fits.[1] We carried out a large number of fits, (around forty thousand), on simulated data randomly deviated around theoretical S-curves and covering a variety of time spans across the width of the life cycle. We fitted the scattered data using Equation 2 from Appendix A with the constant b replaced by at_o so that t_o has the units of time. Each fit yielded values for the three parameters M, a, and t_o. With many fits for every set of conditions, we were able to make distributions of the fitted values and compare their average to the theoretical value used in generating the scattered data. The width of the distributions allowed us to estimate the errors in terms of confidence levels.

The results of our study showed that the more precise the data and the bigger the section of the S-curve they cover, the more accurately the parameters can be recovered. I give below three representative look-up tables. From Table II we can see that historical data covering the first half of the S-curve with 10 percent error per data point will yield a value for the final maximum accurate to within 21 percent, with a 95 percent confidence level.

As an example, let us consider the sales of the US nominal GDP per capita shown at the top of Figure 2.1 in Chapter 2. At the time of the fit, the GDP value was 15,700 and M was estimated as 22,882. Consequently, the S-curve section covered by the data was 15,700/22,882= 68%. From the scattering of the quarterly sales, the statistical error per point, after accounting for seasonal variations, was evaluated as 10 percent. From Table III, then, we obtained the uncertainty on M as around 9 percent for a confidence level of 95 percent. The final ceiling of 8,200 fell within the estimated uncertainty.

Finally, we were able to establish correlations between the uncertainties on the parameters determined. One interesting conclusion, for example, was that among the S-curves that can all fit a set of data, with comparable statistical validity, the curves with smaller values for a have bigger values for M. In other words, a slower rate of growth correlates to a larger niche size and vice-versa. This implies that accelerated growth is associated with a lower ceiling, bringing to mind such folkloric images as short life spans for candles burning at both ends.

TABLE I

Expected uncertainties on M fitted from data covering the range 1 percent to 30 percent of the total S-curve. The confidence level is marked vertically, while the error on the historical data points is marked horizontally. All numbers are in percentages.

	1	5	10	15	20	25
70	2.7	13	28	47	69	120
75	3.2	15	32	53	81	190
80	3.9	17	36	62	110	240
85	4.8	19	41	81	130	370
90	5.9	22	48	110	210	470
95	8.5	29	66	140	350	820
99	48.5	49	180	350	690	

TABLE II

Expected uncertainties on M fitted from data covering the range 1 percent to 50 percent of the total S-curve. The confidence level is marked vertically, while the error on the historical data points is marked horizontally. All numbers are in percentages.

	1	5	10	15	20	25
70	1.2	5.1	11	17	23	23
75	1.4	5.5	12	19	26	32
80	1.8	6.4	14	22	29	36
85	2.1	7.3	16	25	36	42
90	2.6	8.8	18	29	42	48
95	3.1	11.0	21	39	56	66
99	4.6	22.0	30	55	150	110

TABLE III

Expected uncertainties on M fitted from data covering the range 1 percent to 80 percent of the total S-curve. The confidence level is marked vertically, while the error on the historical data points is marked horizontally. All numbers are in percentages.

	1	5	10	15	20	25
70	0.5	1.9	3.9	5.1	8.1	8.9
75	0.6	2.1	4.4	5.5	9.0	9.6
80	0.7	2.4	4.8	6.2	9.8	11.0
85	0.8	2.8	5.5	7.1	12.0	13.0
90	1.1	3.3	6.3	9.1	13.0	16.0
95	1.3	4.0	7.6	11.0	16.0	18.0
99	2.2	5.6	9.1	15.0	21.0	31.0

APPENDIX D

Distinguishing a Logistic from an Exponential

In the solution of the logistic equation (repeated below) M is the value of the final ceiling, t_o is the time of the midpoint and α reflects the steepness of the rising slope.

$$X(t) = \frac{M}{1 + e^{-\alpha(t-t_o)}}$$

It is easy to see that for t large and positive (i.e. $t \gg t_o$) the population $X(t)$ tends to M. Similarly for t large and negative (i.e. $t \ll t_o$) the expression reduces to a simple exponential.

A logistic S-shaped curve is indistinguishable from a simple exponential pattern in its very early stages. There has been controversy about the timing when an S-curve unambiguously distinguishes itself from a simple exponential pattern.[1] Here we present a quantification of this phenomenon.

Let us try to see at what time the S-curve deviates from the exponential pattern in a significant way, see Appendix D Figure 1. Table I quantifies the deviation between a logistic and the corresponding exponential pattern as a fraction of the Logistic's penetration level. By "corresponding" exponential I mean the limit of $X(t)$ as t →-∞.

Divergence of Exponential from S-curve

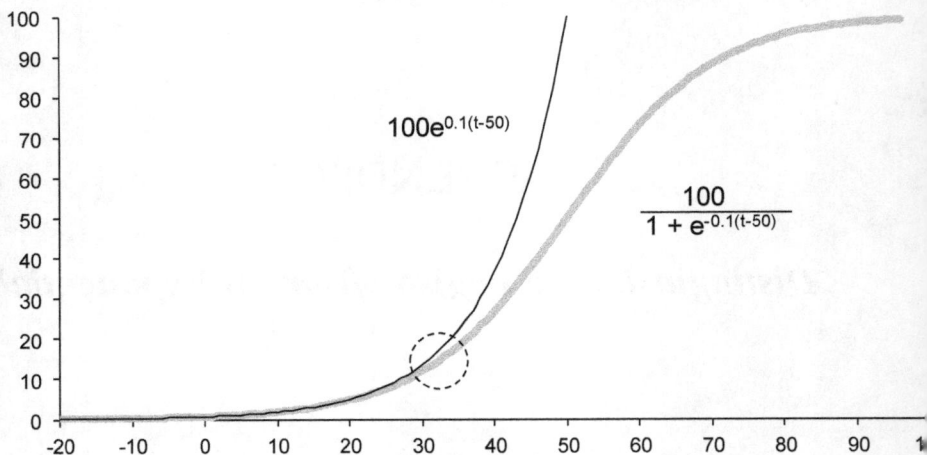

APPENDIX D FIGURE 1 The construction of a theoretical S-curve (gray line) and the exponential (thin black line) it reduces to as time goes backward. The big dotted circle points out the time when the separation becomes unquestionable. The formulae used are shown in the graph.

In Table I we appreciate the size of the deviation between exponential and logistic patterns as a function of how much the logistic has proceeded to completion. Obviously beyond a certain point the difference becomes unquestionable. When exactly this happens maybe subject to judgment so Table I is there to quantitatively help readers make up their mind. Most readers will agree that a 25% deviation between exponential and S-curve patterns is significant because it makes it clear that the two processes can no longer be confused. This happens when the logistic that corresponds to the exponential has reached about 20% of its ceiling level. In other words, the future ceiling that caps a growth process that unquestionably begins deviating from an exponential pattern is about 5 times this level.

TABLE I

The deviation between exponential and logistic patterns as a function of
how much the logistic has proceeded to completion

Deviation	Penetration
11.1%	10.0%
12.2%	10.9%
13.5%	11.9%
15.0%	13.0%
16.5%	14.2%
18.3%	15.4%
20.2%	16.8%
22.3%	18.2%
24.7%	19.8%
27.3%	21.4%
30.1%	23.1%
33.3%	25.0%
36.8%	26.9%
40.7%	28.9%
44.9%	31.0%
49.7%	33.2%
54.9%	35.4%
60.7%	37.8%
67.0%	40.1%
74.1%	42.6%
81.9%	45.0%
90.5%	47.5%
100.0%	50.0%

APPENDIX E

The Canonical Milestones

The dates generally represent an average of clustered events not all of which are mentioned in this table. This is the reason that some events appear misdated, e.g. mass extinction (including dinosaurs) appears too recent by 10 million years. Highlighted in bold is the most outstanding event in the cluster. Present time is taken as year 2000.

No.	Milestone	Years ago
1	**Big bang** and associated processes	1.55×10^{10}
2	**Origin of milky way**/first stars	1.0×10^{10}
3	**Origin of life on Earth**/formation of the solar system and the Earth/oldest rocks	4.0×10^{9}
4	**First eukaryots**/invention of sex (by microorganisms)/atmospheric oxygen/ oldest photosynthetic plants/plate tetonics established	2.1×10^{9}
5	**First multicellular life** (sponges, seaweeds, protozoans)	1.0×10^{9}
6	**Cambrian explosion**/invertebrates/vertebrates/plants colonize land/first reptiles, insects, amphibians	4.3×10^{8}
7	**First mammals**/first birds/first dinosaurs/first use of tools	2.1×10^{8}
8	**First flowering plants**/oldest angiosperm fossil	1.3×10^{8}
9	**Asteroid collision**/first primates/mass extinction (including dinosaurs)	5.5×10^{7}
10	**First humanoids**/first hominids	2.85×10^{7}
11	**First orangutan**/origin of proconsul	1.66×10^{7}
12	**Chimpanzees and humans diverge**/earliest hominid bipedalism	5.1×10^{6}
13	**First stone tools**/first humans/ice age/*homo erectus*/origin of spoken language	2.2×10^{6}
14	**Emergence of *Homo sapiens***	5.55×10^{5}
15	**Domestication of fire**/*Homo heidelbergensis*	3.25×10^{5}
16	**Differentiation of human DNA types**	2.0×10^{5}
17	**Emergence of "modern humans"**/earliest burial of the dead	1.06×10^{5}
18	**Rock art**/protowriting	3.58×10^{4}
19	**Invention of agriculture**	1.92×10^{4}
20	**Techniques for starting fire**/first cities	1.1×10^{4}
21	**Development of the wheel/writinYg**/archaic empires	4,907
22	**Democracy**/city states/the Greeks/Buddha	2,437
23	**Zero and decimals invented**/Rome falls/Moslem conquest	1,440
24	**Renaissance (printing press)**/discovery of new world/the scientific method	539
25	**Industrial revolution (steam engine)**/political revolutions (French, USA)	223
26	**Modern physics**/radio/electricity/automobile/airplane	100
27	**DNA/transistor/nuclear energy**/W.W.II/cold war/sputnik	50
28	**Internet/human genome sequenced**	5

NOTES AND SOURCES

Prologue

1. Theodore Modis, *Predictions*, (New York: Simon & Schuster, 1992).
2. Theodore Modis, *Predictions – 10 Years Later*, (Geneva, Switzerland: Growth Dynamics, 2002).
3. Malcolm Gladwell, *The Tipping Point*, (New York: Little, Brown and Company, 2000).
4. Steven D. Levitt and Stephen J. Dubner. *Freakonomics*, (New York: HarperCollins, 2005).
   http://blogs.smithsonianmag.com/smartnews/2013/02/sorry-malcolm-gladwell-nycs-drop-in-crime-not-due-to-broken-window-theory

Chapter One: The S-Curve

1. J. C. Fisher, and R. H. Pry, "A Simple Substitution Model of Technological Change," *Technological Forecasting and Social Change*, vol. 3, no. 1, (1971): 75-88, and M.J. Cetron and C. Ralph, eds., *Industrial Applications of Technological Forecasting*, (New York: John Wiley & Sons, 1971).
2. T. G. Whiston, "Life Is Logarithmic," in J. Rose, ed., *Advances in Cybernetics and Systems*, (London: Gordon and Breach, 1974).
3. The case was first argued by Cesare Marchetti in "The Automobile in a System Context: The Past 80 Years and the Next 20 Years," *Technological Forecasting and Social Change*, vol. 23 (1983): 3-23. The data for Figure 1.2 come from the *Statistical Abstract of the United States*, US Department of Commerce, Bureau of the Census.
4. John D. Williams, "The Nonsense about Safe Driving," *Fortune*, vol. LVIII, no. 3 (September 1958): 118–19.
5. Ralph Nader, *Unsafe at any Speed*, (New York: Grossman 1965).
6. The data come from Stanley Sadie, ed. *New Grove Dictionary of Music and Musicians*, (London: Macmillan, 1980).

7. The data come from Donald Spoto, *The Dark Side of Genius*, New York: Ballantine Books, 1984).
8. Theodore Modis and Alain Debecker, "Chaoslike States Can Be Observed Before and After Logistic Growth," *Technological Forecasting and Social Change*, vol. 41, (1992): 111-120.

Chapter Two: Limits to Growth

1. Donella H. Meadows, Dennis L. Meadows, Jorgen Randers, William W. Behrens III, *The Limits to Growth*, (New York: Universe Books, New York, 1972).
2. Free forecasts are provided by such organizations as the World Bank, the International Monetary Fund, and the European Commission. Forecasts for a fee are provided by such enterprises as Oxford Economics, Consensus Economics, and the Financial Forecast Center.
3. For forecasts up to 2025 see The Conference Board www.conference-board.org/data/globaloutlook.cfm, and up to 2030 The Economist Intelligence Unit http://www.eiu.com/, US, Economy: long_term outlook, http://country.eiu.com/article.aspx?articleid=1299075914&Country=US&topic=Economy&subtopic=Long-term+outlook&subsubtopic=Summary
4. Tim Jackson, *Prosperity Without Growth*, (London: Earthscan, 2009).
5. Serge Latouche, *Farewell to Growth*, (Malden, MA: Polity Press, 2009).
6. Peter Victor, *Managing Without Growth: Slower by Design, Not Disaster*, (Cheltenham,UK: Edward Elgar, 2008).
7. Richard Heinberg, *The End of Growth*, (Gabriola Island, Canada: New Society Publishers, 2011).
8. This section is based on Theodore Modis, "Long-Term GDP Forecasts and the Prospects for Growth, *Technological Forecasting & Social Change*, vol. 80, (2013): 1557-1562
9. US Department of Commerce, Bureau of Economic Analysis, www.bea.gov/national/index.htm#gdp.
10. A. Debecker, T. Modis, Determination of the Uncertainties of S-curve Logistic Fits, *Technological Forecasting & Social Change*, vol.46, (1994): 153-173.
11. International Monetary Fund, World Economic Outlook: www.imf.org/external/pubs/ft/weo/2012/01/index.htm.
12. Agnus Maddison, *The World Economy*, (Paris: OECD Development Centre Studies, 2004).
13. EconStats: for India http://www.econstats.com/weo/CIND.htm, and for China http://www.econstats.com/weo/CCHN.htm.
14. Data source: BP Statistical Review of World Energy, which can be found at http://www.bp.com/statisticalreview
15. Data source: http://www.census.gov/ipc/www/worldhis.html and http://www.census.gov/popest/
16. Data source is the Nobel Foundation.
17. Data source: US Department of Commerce, Bureau of Economic Analysis and note 12 above.

18. Data source: US Patent and Trademark Office at
    http://www.uspto.gov/web/offices/ac/ido/oeip/taf/issuyear.htm

Chapter Three: Substitutions

1. J. C. Fisher, and R. H. Pry, "A Simple Substitution Model of
   Technological Change," *Technological Forecasting and Social Change*, vol.
   3, no. 1, (1971): 75-88, and M.J. Cetron and C. Ralph, eds., *Industrial
   Applications of Technological Forecasting*, (New York: John Wiley & Sons,
   1971).
2. Steven Schnaars, *Megamistakes: Forecasting and the Myth of Rapid Technological
   Change*, (New York: The Free Press, 1989).
3. Such a graph was originally published by Cesare Marchetti, in
   "Infrastructures for Movement," *Technological Forecasting and Social Change*,
   vol. 32, no. 4 (1987): 373-93, but credit for the original graph must be given
   to Nebosja Nakicenovic, "The Automobile Road to Technological Change:
   Diffusion of the Automobile as a Process of Techological Substitution,"
   Technological Forecasting and Social Change, vol. 29: 309–40.
4. See note 1 above.
5. Adapted from a graph by J. C. Fisher and R. H. Pry in "A Simple
   Substitution Model of Technological Change," *Technological Forecasting and
   Social Change*, vol. 3 (1971): 75-88. Copyright 1988 by Elsevier Science
   Publishing Co., Inc. Reprinted by permission of the publisher.
6. The discussion in this section closely follows ideas first suggested by Cesare
   Marchetti and Nebojsa Nakicenovic at the International Institute of
   Advanced Systems Analysis, Laxenburg, Austria.
7. Nebojsa Nakicenovic, "Software Package for the Logistic Substitution
   Model," report RR-79-12, International Institute of Advanced Systems
   Analysis, Laxenburg, Austria, (1979).
8. The data come from the *Historical Statistics of the United States, Colonial Times to
   1970*, vols. 1 and 2, Bureau of the Census, Washington DC, and from the
   Statistical Abstract of the United States, US Department of Commerce, Bureau
   of the Census.
9. Cesare Marchetti, "The Automobile in a System Context: The Past 80 Years
   and the Next 20 Years," *Technological Forecasting and Social Change*, vol. 23
   (1983): 3-23.
10. Data source: International Civil Aviation Organization, Montreal, Quebec,
    Canada.
11. This figure has been adapted from my article, "Competition and Forecasts
    for Nobel Prize Awards," in *Technological Forecasting and Social Change*, vol. 34
    (1988): 95-102.

Chapter Four: Where Has the Energy Picture Gone Wrong?

1. This chapter draws on the original works by Cesare Marcheti, "Primary Energy Substitution Models: On the Interaction Between Energy and Society", *Technological Forecasting and Social Change*, 10:345—356; Cesare Marchetti, Marchetti, "On Decarbonization: Historically and Perspectively", IR-05-XXX, Prepared for HYDROFORUM 2000, Munich, 11-15 September 2000; and Theodore Modis, "Where Has the Energy Picture Gone Wrong?", *World Future Review*, 1, No 3, June-July 2009.
   Data source: BP Statistical Review of World Energy, which can be found at http://www.bp.com/statisticalreview
2. Data source: http://tonto.eia.doe.gov/
3. Data source: US Energy Information Administration
4. Data sources: Cesare Marcheti, "Infrastructure for Movement", *Technological Forecasting and Social Change*, vol. 32 (1987): 373—393.
   Statistical Review of World Energy 20012 http://www.bp.com/statisticalreview
5. Wood is mostly cellulose that consists of carbon and water. When wood is heated the water gets evaporated and what is left is a black substance 30 percent of which is lignite. These molar ratios have been published in Cesare Marchetti, "When Will Hydrogen Come?" *Int. J. Hydrogen Energy*, 10, 215 (1985).
6. Data sources: *Historical Statistics of the United States, Colonial Times to 1970*, vols. 1 and 2 (Washington DC: Bureau of the Census, 1976) & US Energy Information Administration.
7. Recent data come from the *Statistical Abstract of the United States*, US Department of Commerce, Bureau of the Census.
8. The data used for the S-curve fit have been read off a graph in Jesse H. Ausubel, Cesare Marchetti, and Perrin Meyer, "Toward green mobility: the evolution of transport", originally published in *European Review*, Vol. 6, No. 2, (1998): 137-156.

Chapter Five: Deviations from Natural Growth

1. Theodore Modis and Alain Debecker, "Chaoslike States Can Be Observed Before and After Logistic Growth," *Technological Forecasting and Social Change*, vol. 41, no. 2 (1992).
2. John Naisbitt, *Megatrends*, (New York: Warner Books, 1982).
3. Theodore Modis and Alain Debecker, "Chaoslike States Can Be Expected Before and After Logistic Growth", *Technological Forecasting & Social Change*, vol.41 (1992): 111-120.
4. This graph and most of the discussion in this section come from Theodore Modis, "The normal, the natural, and the harmonic", *Technological Forecasting & Social Change, vol.* 74 (2007): 391-404.

5. Cesare Marchetti, La Saga dei Nobel, *Technol. Rev.* 13 (1989) 8–11 (Italian edition).

6. Theodocre Modis, *Predictions*, (New York: Simon & Schuster, 1992); Theodore Modis, "Competition and Forecasts for Nobel Prize Awards", *Technological Forecasting & Social Change*, vol. 34 (1988): 95-102.

7. Theodore Modis, *Predictions – 10 Years Later*, (Geneva, Switzerland: Growth Dynamics, 2002).

8. B.L. Golden, and P.F. Zantek, "Inaccurate forecasts of the logistic growth model for Nobel prizes", *Technological Forecasting & Social Change*, vol. 71 (2004): 417–422.

9. Theodore Modis, "US Nobel Laureates: Logistic Growth versus Volterra-Lotka", *Technological Forecasting & Social Change*, vol. 78 (2011): 559–564.

Chapter Six: The Kondratieff Cycle

1. Theodore Modis, *Conquering Uncertainty*, (New York: McGraw-Hill, 1996).

2. Nikolai D. Kondratieff, "The Long Wave in Economic Life," *The Review of Economic Statistics*, vol. 17 (1935):105–115.

3. Joseph A. Schumpeter, *Business Cycles* (New York: McGraw-Hill, 1939).

5. See note 2 above.

6. Data source: the Tropical Prediction Center Best Track Reanalysis. http://weather.unisys.com/hurricane/atlantic/.

7. The data have been obtained from P. Kuiper, ed., *The Sun* (Chicago: The University of Chicago Press, 1953) and from P. Bakouline, E. Kononovitch, and V. Moroz, *Astronomie Générale*. Translated into French by V. Polonski. (Moscow: Editions MIR, 1981).
Also from the Internet:
ftp://ftp.ngdc.noaa.gov/STP/SOLAR_DATA/SUNSPOT_NUMBERS/

8. Data source: http://climate.umn.edu/doc/twin_cities/twin_cities.htm

9. Theodore Modis, *Predictions*, (New York: Simon & Schuster, 1992).

10. Nels Winkles III and Iben Browning, *Climate and the Affairs of Men*, (Burlington, VT: Frases, 1975).

11. Such an approach was first published in my book *Conquering Uncertainty*, see note 1 above.

12. W. Brian Arthur, "Positive Feedback in the Economy," *Scientific American*, February 1990, pp 80-85.

13. Theodore Modis, "Fractal Aspects of Natural Growth," *Technological Forecasting & Social Change*, vol. 47, (1994): 63-73.

Chapter Seven: Cascades of S-Curves

1. Theodore Modis, "Fractal Aspects of Natural Growth," *Technological Forecasting & Social Change*, vol. 47, (1994): 63-73.

2. Theodore Modis, "The normal, the natural, and the harmonic", *Technological Forecasting & Social Change*, vol. 74 (2007): 391-404.
3. This figure has been published in the article of note 3 above.
4. Stanley Davis and Bill Davidson, *Vision 2020*, (New York: Simon & Schuster, 1991).
5. See note 1 above.
6. Theodore Modis, *Predictions*, (New York: Simon & Schuster, 1992): 178-179.
7. See note 1 above.
8. Theodore Modis, Why the Singularity Cannot Happen, in *Singularity Hypotheses*, A. H. Eden et al. (eds.), The Frontiers Collection, Springer-Verlag, Berlin Heidelberg, (2012): 311-339.
9. See Note 8 above.
10. See also Note 1 above. The data come from: Laffont, A., and Durieux, F., *Encyclopédie Médico-Chirurgical*, Editions Techniques, Paris, 1985; Kaufmann, Lang, and Rieben, *Croissance de la taille et du poid de 4 à 19.5 ans — Garcons et filles suisses domiciliés dans le canton de Genève en 1972*, Editions Médicine et Hygiène, Geneva, 1976.

Chapter Eight: Genetic Reengineering of Corporations

1. These definitions and subsequent discussion originate in C. Farrell, "Survival of the Fittest Technologies," *New Scientist*, vol. 137, (1993): p. 35; see also C. W. I. Pistorius and J. M. Utterback, "The Death Knells of Mature Technologies", *Technological Forecasting and Social Change*, vol. 50, (1995): 133-151.
2. Peter F. Drucker, "The Discipline of Innovation," *Harvard Business Review*, May-June, (1985): 67-72.
3. Richard N. Foster, *Innovation: The Attacker's Advantage*, (New York: Summit Books, 1986).
4. R. G. Cooper and E. J. Kleinschmidt,: *New Products: The Key Factors in Success*, (Chicago: American Marketing Association, 1990).
5. Kristina Smitalova and Stefan Sujan, *A Mathematical Treatment of Dynamical Models in Biological Science*, (West Sussex, UK: Ellis Horwood, 1991). Actually, credit for the original classification must be given to E. Odum, *Fundamentals of Ecology*, (London: Saunders (1971); and M. Williamson, *The Analysis of Biological Populations,* London: Edward Arnold, 1972).
6. The data come from C. Farrell, "A Theory of Technological Progress," *Technological Forecasting & Social Change*, vol. 44, (1993): p. 161.
7. Frank Sulloway, *Born to Rebel*, (Boston: Pantheon, Harvard University Press, 1996).
8. The truly independent variables are: attractiveness, time constant, and occupancy, the last two as defined below. They are related to but different from the parameters accessible to change. Consequently, some parameters will change together.

The *time constant* of the multiplication process is defined as:

time constant = 1 / log(*attractiveness*)

The *occupancy* is defined as:

occupancy = 1 / [(*time constant*) x (*niche size*)]

See Appendix A for further technical details on how the parameters are related between them.

8. This description has been taken from Farrell, "Theory of Technological Progress," see note 3 above.

9. The data come from Farrell, "Theory of Technological Progress," see note 3 above. Missing data points have been interpolated.

10. Theodore Modis, "US Nobel Laureates: Logistic Growth versus Volterra-Lotka", *Technological Forecasting & Social Change*, vol. 78 (2011): 559–564.

Chapter Nine: The Primordial S-Curve and the Singularity

1. John Brockman, *The Greatest Inventions Of The Past 2, 000 Years*, (New York: Simon & Schuster, 2000).

2. J.W. Schopf (Ed.), *Major Events in the History of Life*, (Boston: Jones and Bartlett Publishers, 1991).
J.D. Barrow, J. Sillk, "The structure of the early universe", *Scientific American* 242 (4) (1980) 118 – 128 (April).

3. Carl Sagan, *The Dragons of Eden: Speculations on the Evolution of Human Intelligence*, (New York: Ballantine Books, 1989).

4. Ray Kurzweil, *The Singularity Is Near*, (New York: Penguin Books, 2005).

5. Theodore Modis, "Forecasting the Growth of Complexity and Change," *Technological Forecasting & Social Change*, vol. 69, (2002): 377 – 404.
Theodore Modis, "The Limits of Complexity and Change," *The Futurist*, May-June, (2003): 26-32.

6. Theodore Modis, "The Singularity Myth," *Technological Forecasting & Social Change*, vol. 73, (2006): 104-112.
Theodore Modis, Why the Singularity Cannot Happen, in *Singularity Hypotheses*, A. H. Eden et al. (eds.), The Frontiers Collection, Springer-Verlag, Berlin Heidelberg, (2012): 311-339.

7. Theodore Modis, "The normal, the natural, and the harmonic", *Technological Forecasting & Social Change, vol.* 74 (2007): 391-404.

8. Theodore Modis and Alain Debecker, "Chaoslike States Can Be Expected Before and After Logistic Growth", *Technological Forecasting & Social Change*, vol.41 (1992): 111-120.

9. Data source: US Energy Information Administration (EIA).

10. Data sources: Intel and Wikipedia.

11. Data sources: United Nations Department of Economic and Social Affairs (UN DESA) and US Census Bureau.

12. Data source: US Department of Commerce, Bureau of Economic Analysis.

13. Theodore Modis, "The End of the Internet Rush," *Technological Forecasting & Social Change*, vol. 72, (2005): 938 – 943.

14. Theodore Modis, "The Normal, the Natural, and the Harmonic," *Technological Forecasting & Social Change*, vol. 74, (2007): 391-404.
15. This figure is from Theodore Modis, *Predictions – 10 Years Later*, (Geneva, Switzerland: Growth Dynamics, 2002). The original figure in *Predictions* had the ceiling of the S-curve at 1% but the data were restated and by 2002 the same data range gave a ceiling of 2.3%.
    The data come from HIV/AIDS Surveillance, Centers for Disease Control, US Department of Health and Human Services, Atlanta, GA.
16. Alain Debecker and Theodore Modis, "Determination of the Uncertainties in S-curve Logistic Fits," *Technological Forecasting & Social Change*, vol. 46 (1994): 153-173.
17. Cesare Marchetti, "10¹²: A Check on the Earth-Carrying Capacity for Man," *Energy*, vol. 4, (1979):1107—1117.
18. Data source: United Nations Population Division (U.N. 1999).
19. http://bitcoin.org/bitcoin.pdf

Epilogue

1. Alain Debecker and Theodore Modis, "Determination of the Uncertainties in S-curve Logistic Fits," *Technological Forecasting & Social Change*, vol. 46 (1994): 153-173.
2. K. Axelos, *Héraclite et la philosophie*, (Paris: Les Éditions de Minuit, 1962).

Appendix B: Mathematical Formulations of S-Curves and the Procedure for Fitting Them on Data

1. E. W. Montroll and N. S. Goel, "On the Volterra and Other Nonlinear Models of Interacting Populations," *Review of Modern Physics*, vol. 43, no. 2 (1971): 231.
2. M. Peschel and W. Mendel, *Leben wir in einer Volterra Welt?* (Berlin: Akademie Verlag, 1983).
3. See note 1 above.
4. P. F. Verhulst, "Recherches mathématiques sur la loi d'Accroissement de la Population" ("Mathemtical Research on the Law of Population Growth"), *Nouveaux Memoires de l'Académie Royale des Sciences et des Belles-Lettres de Bruxelles*, vol. 18 (1945): 1-40; also in P. F. Verhulst, "Notice sur la loi que la population suit dans son accroissement" (Announcement on the Law Followed by a Population During Its Growth"), *Correspondence Mathématique et physique*, vol. 10: 113–21.
5. J. B. S. Haldane, "The Mathematical Theory of Natural and Artificial Selection," *Transactions, Cambridge Philosophical Society*, vol. 23 (1924): 19-41.
6. Alfred J. Lotka, *Elements of Physical Biology*, (Baltimore: Williams & Wilkins Co., 1925).

7. Theodore Modis, *Natural Laws in the Service of the Decision Maker*, (Lugano, Switzerland: Growth Dynamics, 2013); or Theodore Modis, *Decision-Making for a New World*, (Frankfurt/New York: Campus Verlag, 2014).

8. Theodore Modis, "The Normal, the Natural, and the Harmonic", *Technological Forecasting & Social Change*, vol. 74 (2007): 391-404.

9. Theodore Modis and Alain Debecker, "Chaoslike States Can Be Observed Before and After Logistic Growth," *Technological Forecasting and Social Change*, vol. 41, (1992): 111-120.

Appendix C: Expected Uncertainties on S-Curve Forecasts

1. Alain Debecker and Theodore Modis, "Determination of the Uncertainties in S-curve Logistic Fits," *Technological Forecasting & Social Change*, vol. 46 (1994): 153-173.

Appendix D: Distinguishing a Logistic from an Exponential

1. Supporters of the Singularity notion have been dealing with trends they call exponential but which in fact are S-curves, as discussed in Chapter 9.

ACKNOWLEDGEMENTS

I am indebted to the work of various scientists who have worked with S-curves during the last 150 years and whose names are cited throughout Notes and Sources. These people have shared a passion for the same subject that has become my life's work. Among them, Cesare Marchetti requires special mention. He has contributed enormously to making the formulation of natural growth a general vehicle for understanding society. I have drawn extensively on his ideas, ranging from the concept of invariants to the productivity/creativity of artists and scientists and the evolution of nuclear energy. I present many of his results and carry further his analyses. He has masterminded the explanation of natural decarbonization and the intricate interdependence of primary-energy systems and transport infrastructures discussed in Chapter 4. This book could not have come into existence had it not been for the work of Cesare Marchetti and his collaborators, Nebojsa Nakicenovic and Arnulf Grubler, at the International Institute for Applied Systems Analysis in Laxenburg, Austria. I am grateful to them for keeping me abreast of their latest work and for all the fruitful discussions we have had together.

Of great importance has been the contribution of my collaborator and friend, Alain Debecker. His personal and professional support reflects on many aspects of this book. I finally want to thank my colleague and friend of old times Eric Schwartz whose idea was to make a graph of the most important milestones in the evolution of the universe. Our many exchanges largely shaped Chapter 9.

Theodore Modis
Lugano, Switzerland

December 2013

INDEX

Page numbers in **bold** refer to principal discussions
Page number in *italics* refer to figures

ABOUT THE AUTHOR

Theodore Modis studied electrical engineering (Bachelor, Masters) and subsequently physics (Ph.D.) at Columbia University in New York. He worked for 15 years as researcher in high-energy physics experiments at Brookhaven National Laboratories and Europe's CERN. In 1984 he joined Digital Equipment Corporation where he led a group of management-science consultants. In 1994 he founded Growth Dynamics, (www.growth-dynamics.com) an organization specializing in strategic forecasting and management consulting.

He is author/co-author to about 100 articles in scientific and business journals, and several software programs including iPhone/iPad applications that bring the S-curves to the stock market. He has also written nine books that have been translated in several languages:

- *Predictions: Society's Telltale Signature Reveals the Past and Forecasts the Future*, Simon & Schuster, New York, 1992.

- *Conquering Uncertainty: Understanding Corporate Cycles and Positioning Your Company to Survive the Changing Environment*, McGraw-Hill, New York, 1998.

- *An S-Shaped Trail to Wall Street: Survival of the Fittest Reigns in the Stock Market*, Growth Dynamics, Geneva, 1999.

- *Predictions: 10 Years Later*, Growth Dynamics, Geneva, 2002.

- *Street Science: A Physicist's Wanderings off the Beaten Track*, Growth Dynamics, Lugano, Switzerland, 2009.

- *Bestseller Driven: A Book-Writing Story*, Growth Dynamics, Lugano, Switzerland, 2005.

- *Natural Laws in the Service of the Decision Maker: How to Use Science-Based Methodologies to See More Clearly further into the Future*, Growth Dynamics, Lugano, Switzerland, 2013.

- *Decision-Making for a New World: Natural Laws of Evolution and Competition as a Road Map to Revolutionary New Management*, Campus Verlag – editionMALIK, New York, 2014.
- *An S-Shaped Adventure: Predictions – 20 Years Later*, Growth Dynamics, Lugano, Switzerland, 2014.

He has taught at Columbia University, the University of Geneva, and at the European business schools INSEAD and IMD. He has been a member of the faculty of DUXX Graduate School in Business Leadership, in Monterrey, Mexico. He lives in Lugano, Switzerland.

www.ingramcontent.com/pod-product-compliance
Lightning Source LLC
Chambersburg PA
CBHW030836300326
41935CB00036B/178